Good Questions

Great Ways to Differentiate Mathematics Instruction in the Standards-Based Classroom

Third Edition

Also by Marian Small

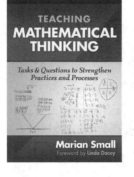

Teaching Mathematical Thinking:
Tasks and Questions to Strengthen
Practices and Processes

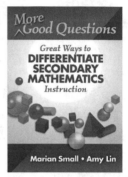

More Good Questions:
Great Ways to Differentiate
Secondary Mathematics Instruction
(with Amy Lin)

Building Proportional Reasoning
Across Grades and
Math Strands, K–8

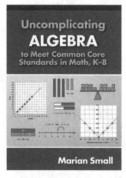

Uncomplicating Algebra
to Meet Common Core
Standards in Math, K–8

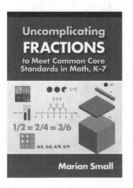

Uncomplicating Fractions
to Meet Common Core
Standards in Math, K–7

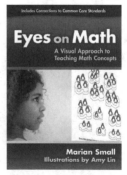

Eyes on Math:
A Visual Approach to
Teaching Math Concepts
(Illustrations by Amy Lin)

Good Questions

Great Ways to Differentiate Mathematics Instruction in the Standards-Based Classroom

Third Edition

MARIAN SMALL

Foreword *by Carol Ann Tomlinson*

TEACHERS COLLEGE PRESS

TEACHERS COLLEGE | COLUMBIA UNIVERSITY
NEW YORK AND LONDON

NATIONAL COUNCIL OF
TEACHERS OF MATHEMATICS
1906 Association Drive, Reston, VA 20191
www.nctm.org

NELSON
nelson.com

Published simultaneously by Teachers College Press, 1234 Amsterdam Avenue, New York, NY 10027, and National Council of Teachers of Mathematics, 1906 Association Drive, Reston, VA 20191; distributed in Canada by Nelson Education, 1120 Birchmount Road, Toronto, ON, Canada M1K 5G4.

Credits: Teddy bear, boat, and frame (page 148) from *Nelson Math Focus 2* by Marian Small, page 74. Copyright © 2008. Reprinted with permission of Nelson Education Limited.

Text Design: Lynne Frost

Library of Congress Cataloging-in-Publication Data

Names: Small, Marian.
Title: Good questions : great ways to differentiate mathematics instruction in the
    standards-based classroom / Marian Small ; foreword by Carol Ann Tomlinson.
Description: Third edition. | New York, NY : Teachers College Press, [2017] | Subtitle
    varies slightly from previous edition. | Includes bibliographical references and index.
Identifiers: LCCN 2017007400 (print) | LCCN 2017008595 (ebook) |
    ISBN 9780807758540 (pbk. : alk. paper) | ISBN 9780807775851 (ebook)
Subjects: LCSH: Mathematics—Study and teaching (Elementary) | Individualized
    instruction. | Effective teaching.
Classification: LCC QA20.I53 S63 2017 (print) | LCC QA20.I53 (ebook) |
    DDC 372.7—dc23
LC record available at https://lccn.loc.gov/2017007400

ISBN 978-0-8077-5854-0 (paper)
ISBN 978-0-8077-7585-1 (ebook)
NCTM Stock Number 15474

Printed on acid-free paper
Manufactured in the United States of America

25 24 23 22 21 19 18 17   8 7 6 5 4 3 2 1

Contents

Foreword

THAT I WOULD BE writing a Foreword for a book on teaching math seems at once both ironic and absolutely appropriate.

The irony stems from my long and generally unhappy life as a math student. In elementary school, math was neither easy nor hard for me. It was simply something I did. It evoked neither pleasure nor pain. I suppose it was a bit like household chores—something necessary to get through the day, but certainly nothing to be coveted. Through those years, math was as rote as household chores. I memorized the required number sets, counting, and multiplication, absorbed algorithms I watched my teacher scribe on the blackboard in the front of the room, and repeated what I saw. There was no joy in it, but it was doable.

Once I entered the world of algebra, however, math became a sinister thing. Letters invaded the numbers. Equal signs took on a super power. While I could occasionally duplicate what my teacher chalked in front of us, I could no longer commit the strings to memory. There was no reason to them, and worse, I didn't even really grasp the notion that they were all about reasoning. I simply knew that I could no longer succeed with math. The conclusion I drew, of course, was that I was no good at math.

I received a one-year reprieve from my self-imposed verdict when geometry intervened between Algebra I and Algebra II—and before trigonometry had its way with me. In that one year, math was glory. In that one year, math was about reasoning. There was order to it, like the order in the universe. There were words that marched along with the figures and problems. I loved it in the way that I loved good writing. It revealed beauty and it dignified my possibilities as a learner rather than eroding them.

But the interlude didn't last long enough to erase the damage done earlier or to reconfigure my sense of myself as outpriced by math. It took many years for me to realize that the dismay I felt in the presence of math was neither a fault in me nor an indictment of math as a content area. Rather, I came to see myself as one among a legion of students who had abandoned a content area—or even school—because the way a subject was taught drained all the life from it, even as it drained life from the learners.

The appropriateness of my writing this Foreword is, of course, not completely separate from the irony. I have spent my career as a teacher (an English teacher, to be clear) and then as a professor with an evolving belief, then a conviction, and finally a passion for teaching that opens up possibilities and a life for every student in a classroom. In time, I applied a name to the system of ideas that pointed the way to achieving that sort of success in academically diverse classrooms. That name, of course, is *differentiation*.

I have read and continue to read many books related to differentiation. I learn from many of them, and I turn away from a good share of them as well, when they appear to trivialize differentiation or minimize the complexity of teaching a broad range of students in a meaningful way.

In Marian's work, I find an educational soulmate.

The two strategies she presents in this book are streamlined—they even seem spare. And yet in her writing about them, I see all of the components I have come to value in classrooms that serve academically diverse learners in a powerful way. The five elements I write about as core to effective differentiation are: an invitational learning environment, a meaning-rich curriculum, assessment that informs both teaching and learning, consistent attention to learner differences through responsive instruction, and creating classroom routines that balance predictability and flexibility.

The focus of this book is on meaning-rich curriculum and responsive instruction through open questions and parallel tasks. In both areas, I not only see reflected the best of what our field knows about curriculum and teaching in response to students' varied entry points, but I see what math should always be—should always have been—to students: a language of reason, a catalyst for developing every learner's capacity to think, a way forward in the world. This book lays out those elements through explanation and example.

Woven more subtly, but no less powerfully, throughout the book are the other elements I've come to champion as well. When a teacher asks questions that every student can contribute to answering, and presents work that provides options for student exploration, the learning environment becomes more invitational for every learner. The atmosphere of winners and losers is diminished. Community develops.

The book speaks, too, to the reality that the teacher who consistently observes and listens to students as they work derives insights that contribute to more informed and effective teaching. And in a very interesting way, the use of open questions and parallel tasks (rather than adherence to fixed and coverage-focused routines) requires teachers to become more focused on supporting success for a range of students.

Freeman Hrabowski, a noted mathematician and president of the University of Maryland Baltimore County, says that math gives him goose bumps, and always has. If I almost never understood that reaction as a student of math, I do now when I read Marian Small's work. Not only do I see the power of math revealed, but I also see her work opening that power to so many students who might never otherwise experience it—and to teachers whose own math experiences, like mine, were one-dimensional at the very best.

I am a disciple of Marian Small's approach to differentiating math!

—Carol Ann Tomlinson, Ed.D.
William Clay Parrish, Jr. Professor & Chair
Educational Leadership, Foundations, & Policy
Curry School of Education, University of Virginia

Preface

FOR THE PAST 10 or more years, I have had the opportunity to develop a number of resources that support teachers who seek to differentiate instruction in mathematics. Much of the work I do focuses on differentiation in terms of student readiness and highlights two strategies: open questions and parallel tasks.

These strategies have been employed effectively by thousands of teachers throughout Canada and the United States, as well as overseas. What we have learned is that to differentiate instruction in math, it is important for teachers to think about "bigger ideas," as they do when teaching other subjects, rather than focus intensely on the usual very tight content pieces in mathematics that capture their attention.

ORGANIZATION OF THE BOOK

An introductory chapter of this resource describes the rationale for differentiating math instruction and explains the two principal strategies that are employed throughout the book: open questions and parallel tasks. Nine content chapters, built around the organization of the content standards in the Common Core State Standards for Mathematics (Common Core Standards Initiative, 2010) then illustrate applications of these strategies, followed by a final concluding chapter, an appendix listing the Common Core content domains and mathematical practices, an appendix containing a template for teachers wishing to develop their own materials, a glossary, a bibliography, and an index.

Chapters 1–9 focus on the following content topics described in the Common Core Standards for Mathematics.

- Counting & Cardinality / Number & Operations in Base Ten
- Number & Operations—Fractions
- Ratios & Proportional Relationships
- The Number System
- Operations & Algebraic Thinking
- Expressions & Equations / Functions
- Measurement & Data
- Geometry
- Statistics & Probability

The content strands are not developed sequentially, so Chapters 1–9 can be approached in any order. Within each of the content chapters, a set of big ideas is described that underlies the increasingly sophisticated teaching of that content throughout the grades. In this edition, each question or task is tagged to one or more of those big ideas, as well as to applicable Common Core Standards for Mathematical Practice.

The suggested differentiating questions and tasks are organized according to the primary, elementary, and middle-level grade bands set out in the Common Core Standards. The object of differentiation is to teach the same broad concepts to students at different developmental levels. The multilayered organization of material within Chapters 1–9 is intended to help make this possible.

Appendix A provides a listing of the Common Core content domains keyed to chapter and grade band coverage in this resource, as well as a listing of the Common Core Mathematical Practice Standards.

Appendix B features a template worksheet that will assist teachers in developing their own materials in support of differentiated instruction through use of open questions and parallel tasks. An example of application of the worksheet appears in the Introduction.

The Glossary defines technical terms used throughout. Each word that appears in the Glossary is shown in boldface type at its first occurrence in the text, and each Glossary entry is annotated with the chapter and page number of the term's first occurrence.

The Bibliography highlights three types of resources: those that are referenced in other parts of the text, those that speak to the issues of teaching to big ideas and differentiating instruction, as well as a number of excellent sources for activities that can be used as-is or used as a starting point for creating open questions and parallel tasks.

The Index focuses on educational concepts—standards, student development, teaching methods and principles, and such—as opposed to mathematical concepts. To facilitate user access to the mathematical topics covered, an Index of Big Ideas is provided, listing all big ideas covered in the content chapters.

CHANGES IN THE THIRD EDITION

Readers familiar with the second edition of this book will note that there are many new questions (approximately 30%) and some rewording of the big ideas that are described in each chapter. The biggest change, though, is the use of the Common Core organization for the content in order to make it easier for teachers following the Common Core or similar standards to locate questions more easily.

The eight Common Core Standards for Mathematical Practice that were addressed in the second edition are also addressed in this edition. These practice standards (which are listed in Appendix A) work hand-in-hand with the CCSS content domain standards. They reflect "processes" that should be elicited from students and goals for orienting and developing students as mathematical thinkers as content is learned.

Some of these practices may or may not be brought out, depending on how a teacher handles the presented questions or tasks, but others are implicit no matter what direction the discussion of the question or task takes.

For example, almost all of the questions and tasks posed in this book require students to make sense of them and persevere in solving them. This is true with open questions, which are often deliberately vague and require students to make sense of the question before they can choose the direction in which to go. This is also true with parallel tasks, where students must make sense of both options to decide with which one to proceed.

Students are frequently asked to explain their rationale, requiring them to construct viable arguments. Ideally, teachers would encourage students to critique each other's reasoning, although that would not be implicit in the question itself.

A great many of the questions provided require students to reason, either abstractly or, in the case of number or measurement questions, quantitatively.

Some questions bring out the other practices: modeling, using appropriate tools strategically, attending to precision, looking for and making use of structure, and/or looking for and expressing regularity in repeated reasoning. There are fewer questions attending to precision in this particular resource than in others precisely because differentiation must allow for varying levels of precision.

A focus on expressing regularity in repeated reasoning is also less prominent in this particular resource because the questions are frequently intended to generate initial discussion rather than to tie down rules. There are a few examples, however, that do have that focus.

Listed with each question and task are those Common Core Mathematical Practices that are clearly evoked in the question or task. This does not mean that teachers might not also evoke others, depending on their approach to student responses or the student responses themselves.

IT IS MY HOPE that teachers will embrace the two core strategies—open questions and parallel tasks—that are explained and demonstrated in this book, and find them, as I have, to be helpful to the many children who come into classrooms with highly differentiated mathematical preparation, skill, and confidence. Seeing a child who has been struggling in mathematics start to feel successful is an important goal in my teaching. I have seen the use of the strategies described in this volume make that happen over and over again.

Acknowledgments

THIS BOOK has provided me with an opportunity to respond to the many kind teachers who tell me how much the early editions of this book have helped them change the climate in their math classrooms for the better. I hope that a new organization and more questions will add to that support.

Again, I wish to thank Teachers College Press and, in particular, Jean Ward, Lynne Frost, and Karl Nyberg for being so easy to work with and so positive toward my work.

Finally, I would like to thank Carol Ann Tomlinson for her lovely Foreword.

Introduction

Why and How to Differentiate Math Instruction

STUDENTS IN ANY CLASSROOM differ in many ways, only some of which the teacher can reasonably attend to in developing instructional plans. Some differences will be cognitive—for example, what previous concepts and skills students can call upon. Some will be more about learning style and preferences, including behaviors such as persistence or inquisitiveness or the lack thereof; whether the student learns better through auditory, visual, or kinesthetic approaches; and personal interests.

THE CHALLENGE IN MATH CLASSROOMS

Although many teachers of language arts recognize that different students need different reading material, depending on their reading level, it is often more challenging for teachers to vary the material they ask their students to work with in mathematics. The math teacher will more frequently teach all students based on a fairly narrow curriculum goal presented in a textbook, even though teachers have been asked to encourage a variety of strategies in the solution of problems. The teacher will recognize that some students need additional help and will provide as much support as possible to those students while the other students are working independently. **Differentiating instruction** in mathematics is still a relatively new idea. It is not easy in mathematics to simply provide an alternate book to read (as can be done in language arts). And a majority of teachers have never been trained to really understand how students differ mathematically. However, students in the same grade level clearly *do* differ mathematically in significant ways. Teachers want to be successful in their instruction of all students, and feel even more pressure to do so in the current social climate. Understanding differences and differentiating instruction are important processes for achievement of that goal.

The National Council of Teachers of Mathematics (NCTM), the professional organization whose mission it is to promote, articulate, and support the best possible teaching and learning in mathematics, recognizes the need for differentiation. The first principle of the NCTM *Principles and Standards for School Mathematics* reads, "Excellence in mathematics education requires equity—high expectations and strong support for all students" (NCTM, 2000, p. 12).

In particular, NCTM recognizes the need for accommodating differences among students, taking into account prior knowledge and intellectual strengths, to ensure that each student can learn important mathematics. "Equity does not mean that every student should receive identical instruction; instead, it demands that reasonable and appropriate accommodations be made as needed to promote access and attainment for all students" (NCTM, 2000, p. 12). This has been addressed more recently, after the widespread adoption of the Common Core Standards, in *Principles to Action* (NCTM, 2014), where a contrast is made between unproductive beliefs about access and equity in mathematics and more productive beliefs.

How Students Might Differ

One way that we see the differences in students is through their responses to the mathematical questions and problems that are put to them. For example, consider the task below, which might be asked of 3rd-grade students:

> In one cupboard, you have three shelves with five boxes on each shelf. There are three of those cupboards in the room. How many boxes are stored in all three cupboards?

Students might approach the task in very different ways. Here are some examples:

- Liam immediately raises his hand and simply waits for the teacher to help him.
- Angelita draws a picture of the cupboards, the shelves, and the boxes and counts each box.
- Tara uses addition and writes $5 + 5 + 5 + 5 + 5 + 5 + 5 + 5 + 5$.
- Dejohn uses addition and writes $5 + 5 + 5 = 15$, then adds again, writing $15 + 15 + 15 = 45$.
- Rebecca uses a combination of multiplication and addition and writes $3 \times 5 = 15$, then $15 + 15 + 15 = 45$.

The Teacher's Response

What do all these different student approaches mean for the teacher? They demonstrate that quite different forms of feedback from the teacher are needed to support the individual students. For example, the teacher might wish to:

- Follow up with Tara and Dejohn by introducing the benefits of using a multiplication expression to record their thinking.

- Help Rebecca extend what she already knows about multiplication to more situations.
- Encourage Liam to be more independent, or set out a problem that is more suitable to his developmental level.
- Open Angelita up to the value of using more sophisticated strategies by setting out a problem in which counting becomes even more cumbersome.

These differences in student approaches and appropriate feedback underscore the need for a teacher to know where his or her students are developmentally to be able to meet each one's educational needs. The goal is to remove barriers to learning while still challenging each student to take risks and responsibility for learning (Karp & Howell, 2004).

WHAT IT MEANS TO MEET STUDENT NEEDS

One approach to meeting each student's needs is to provide tasks within each student's **zone of proximal development** and at the same time to ensure that each student in the class has the opportunity to make a meaningful contribution to the class community of learners. Zone of proximal development is a term used to describe the "distance between the actual development level as determined by independent problem solving and the level of potential development as determined through problem solving under adult guidance or in collaboration with more capable peers" (Vygotsky, 1978, p. 86).

Instruction within the zone of proximal development allows students, whether through guidance from the teacher or through working with other students, to access new ideas that are close enough to what they already know to make the access feasible. Teachers are not using educational time optimally if they either are teaching beyond a student's zone of proximal development or are providing instruction on material the student already can handle independently. Although other students in the classroom may be progressing, the student operating outside his or her zone of proximal development is often not benefiting from the instruction.

For example, a teacher might be planning a lesson on multiplying a decimal by a whole number. Although the skill that is the goal of the lesson is to perform a computation such as 3×1.5, there are three underlying mathematical concepts that a teacher would want to ensure that students understand. Students working on this question should know:

- What multiplication means (whether repeated addition, the counting of equal groups, the calculation of the area of a rectangle, or the description of a rate [three times as many])
- That multiplication has those same meanings regardless of what number 3 is multiplying
- That multiplication can be accomplished in parts (the distributive principle), for example, $3 \times 1.5 = 3 \times 1 + 3 \times 0.5$

Although the planned lesson is likely to depend on the fact that students understand that 1.5 is 15 tenths or 1 and 5 tenths, a teacher could effectively teach the same lesson even to students who do not have that understanding or who simply are not ready to deal with decimals. The teacher could allow the less developed students to explore the concepts using whole numbers while the more advanced students are using decimals. Only when the teacher felt that the use of decimals was in an individual student's zone of proximal development would the teacher ask that student to work with decimals. Thus, by making this adjustment, the teacher differentiates the task to locate it within each student's zone of proximal development.

ASSESSING STUDENTS' NEEDS

For a teacher to teach to a student's zone of proximal development, first the teacher must determine what that zone is by gathering diagnostic information to assess the student's mathematical developmental level. For example, to determine a 3rd- or 4th-grade student's developmental level in multiplication, a teacher might use a set of questions to find out whether the student knows various meanings of multiplication, knows to which situations multiplication applies, can solve simple problems involving multiplication, and can multiply single-digit numbers, using either memorized facts or strategies that relate known facts to unknown facts (e.g., knowing that 6×7 must be 7 more than 5×7).

Some tools to accomplish this sort of evaluation are tied to developmental continua that have been established to describe students' mathematical growth (Small, 2005a, 2005b, 2006, 2007, 2010b; Small et al., 2011a, 2011b). Teachers might also use locally or personally developed diagnostic tools. Only after a teacher has determined a student's level of mathematical sophistication, can he or she even begin to attempt to address that student's needs.

PRINCIPLES AND APPROACHES TO DIFFERENTIATING INSTRUCTION

Differentiating instruction is not a new idea, but the issue has been gaining an ever higher profile for mathematics teachers in recent years. More and more, educational systems and parents are expecting the teacher to be aware of what each individual student needs and to plan instruction to focus on those needs. In the past, this was less the case in mathematics than in other subject areas, but now the expectation is common in mathematics as well.

There is general agreement that to effectively differentiate instruction, the following elements are needed:

- *Big Ideas.* The focus of instruction must be on the **big ideas** being taught to ensure that they all are addressed, no matter at what level.
- *Choice.* There must be some aspect of choice for the student, whether in content, process, or product.

- *Preassessment.* Prior assessment is essential to determine what needs different students have (Gregory & Chapman, 2006; Murray & Jorgensen, 2007).

Teaching to Big Ideas

The Curriculum Principle of the NCTM *Principles and Standards for School Mathematics* states that "A curriculum is more than a collection of activities: it must be coherent, focused on important mathematics, and well articulated across the grades" (NCTM, 2000, p. 14). The introduction to the Common Core State Standards indicates that the Standards not only stress conceptual understanding of key ideas but also continually return to organizing principles, such as properties of operations, to structure those ideas (Common Core State Standards Initiative, 2010).

Curriculum coherence requires a focus on interconnections, or big ideas. Big ideas represent fundamental principles; they are the ideas that link the specifics. For example, the notion that **benchmark numbers** are a way to make sense of other numbers is equally useful for the 1st-grader who relates the number 8 to the more familiar 10, the 4rd-grader who relates $\frac{3}{8}$ to the more familiar $\frac{1}{2}$, or the 7th-grader who relates π to the number 3. If students in a classroom differ in their readiness, it is usually in terms of the specifics and not the big ideas. Although some students in a classroom where rounding of decimal thousandths to appropriate benchmarks is being taught might not be ready for that precise topic, they could still deal with the concept of estimating, when it is appropriate, and why it is useful.

Big ideas can form a framework for thinking about "important mathematics" and supporting standards-driven instruction. Big ideas find application across all grade bands. There may be differences in the complexity of their application, but the big ideas remain constant. Many teachers believe that curriculum requirements limit them to fairly narrow learning goals and feel that they must focus instruction on meeting those specific student outcomes. Differentiation requires a different approach, one that is facilitated by teaching to the big ideas.

Choice

Many math teachers are still not comfortable with the notion of student choice except in the rarest of circumstances. They worry that students will not make "appropriate" choices.

However, some teachers who are uncomfortable differentiating instruction in terms of the main lesson goal are willing to provide some choice in follow-up activities students use to practice the ideas they have been taught. Some of the strategies that have been suggested for differentiating practice include use of menus from which students choose, tiered lessons in which teachers teach to the whole group and vary the follow-up for different students, learning stations where different students attempt different tasks, or other approaches that allow for student choice, usually in pursuit of the same basic overall lesson goal (Tomlinson, 1999; Westphal, 2007).

For example, a teacher might present a lesson on creating equivalent fractions to all students, and then vary the follow-up. Some students might work only with simple fractions and at a very concrete level; these tasks are likely to start with simple fractions where the numerator and denominator have been multiplied (but not divided) to create equivalent fractions. Other students might be asked to work at a pictorial or even symbolic level with a broader range of fractions, where numerators and denominators might be multiplied or divided to create equivalent fractions and more challenging questions are asked (e.g., *Is there an equivalent fraction for $\frac{10}{15}$ where the denominator is 48?*).

Preassessment

To provide good choices, a teacher must first know how students in the classroom vary in their knowledge of facts and in their mathematical developmental level. This requires collecting data either formally or informally to determine what abilities and what deficiencies students have. Although many teachers feel they lack the time or the tools to preassess on a regular basis, the data derived from preassessment are essential in driving differentiated instruction.

Despite the importance of preassessment, employing a highly structured approach or a standardized tool for conducting the assessment is not mandatory. Depending on the topic, a teacher might use a combination of written and oral questions and tasks to determine an appropriate starting point for each student. Sometimes this can be done informally by giving students just a few or even one revealing task or problem. Students can also be asked to rate the difficulty that they had with the problem based on a scale of 1–5, with the teacher making clear to them that the information is only to inform the teacher's instruction, not to evaluate the student's work.

TWO CORE STRATEGIES FOR DIFFERENTIATING MATHEMATICS INSTRUCTION: OPEN QUESTIONS AND PARALLEL TASKS

It is not realistic for a teacher to try to create 30 different instructional paths for 30 students, or even 6 different paths for 6 groups of students. Because this is the perceived alternative to one-size-fits-all teaching, instruction in mathematics is often not differentiated. To differentiate instruction efficiently, teachers need manageable strategies that meet the needs of most of their students at the same time. Through use of just two core strategies, teachers can effectively differentiate instruction to suit most students. These two core strategies underlie all of the questions and tasks in this book:

- **Open questions**
- **Parallel tasks**

Open Questions

The ultimate goal of differentiation is to meet the needs of the varied students in a classroom during instruction. This becomes manageable if the teacher can create a single question or task that is inclusive not only in allowing for different students to approach it by using different processes or strategies but also in allowing for students at different stages of mathematical development to benefit and grow from attention to the task. In other words, the task is in the appropriate zone of proximal development for the entire class. In this way, each student becomes part of the larger learning conversation, an important and valued member of the learning community. Struggling students are less likely to be the passive learners they so often are (Lovin, Kyger, & Allsopp, 2004).

A question is open when it is framed in such a way that a variety of responses or approaches are possible. Consider, for example, these two questions, each of which might be asked of a whole class, and think about how the results for each question would differ:

Question 1:	To which fact family does the fact 3 × 4 = 12 belong?
Question 2:	Describe the picture at the right by using a mathematical equation. x x x x x x x x x x x x

If the student does not know what a fact family is, there is no chance he or she will answer Question 1 correctly. In the case of Question 2, even if the student is not comfortable with multiplication, the question can be answered by using addition statements (e.g., 4 + 4 + 4 = 12 or 4 + 8 = 12). Other students might use multiplication statements (e.g., 3 × 4 = 12 or 4 × 3 = 12), division statements (e.g., 12 ÷ 3 = 4 or 12 ÷ 4 = 3), or even statements that combine operations (e.g., 3 × 2 + 3 × 2 = 12).

A Different Kind of Classroom Conversation. Not only will the mathematical conversation be richer in the case of Question 2—the open question—but almost any student can find something appropriate to contribute.

The important point to notice is that the teacher can put the same question to the entire class, but the question is designed to allow for differentiation of response based on each student's understanding. All students can participate fully and gain from the discussion in the classroom learning community.

This approach differs, in an important way, from asking a question, observing students who do not understand, and then asking a simpler question to which they can respond. By using the open question, students gain confidence; they can answer the teacher's question right from the start. Psychologically, this is a much more positive situation.

Multiple Benefits. There is another benefit to open questions. Many students and many adults view mathematics as a difficult, unwelcoming subject because they see it as black and white. Unlike, for instance, social studies or English, where people might be encouraged to express different points of view, math is viewed as a subject where either you get it or you don't. This view of mathematics inhibits many students from even trying. Once they falter, they assume they will continue to falter and may simply shut down.

It is the job of teachers to help students see that mathematics is multifaceted. Any mathematical concept can be considered from a variety of perspectives, and those multiple perspectives actually enrich its study. Open questions provide the opportunity to demonstrate this.

Strategies for Creating Open Questions. This book illustrates a variety of styles of open questions. Some common strategies that can be used to convert conventional questions to open questions are described below:

- Turning around a question
- Asking for similarities and differences
- Replacing a number with a blank
- Creating a sentence
- Using "soft" words
- Changing the question

Turning Around a Question. For the turn-around strategy, instead of giving the question, the teacher gives the answer and asks for the question. For example:

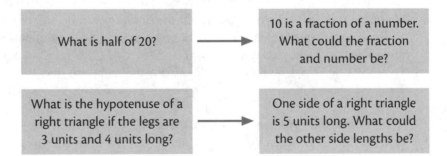

Asking for Similarities and Differences. The teacher chooses two items—two numbers, two shapes, two graphs, two probabilities, two measurements, and so forth—and asks students how they are alike and how they are different. Inevitably, there will be many good answers. For example, the teacher could ask how the number 85 is like the number 100 and how it is different. A student might realize that both numbers are said when skip counting by 5s, both are less than 200, and both are greater than 50, but only one is a three-digit number, only one ends with a 5, and only one is greater than 90.

Replacing a Number with a Blank. Open questions can be created by replacing a number with a blank and allowing the students to choose the number to use. For example, instead of asking how many students there are altogether if there are 25 in one class and 31 in another, students could be asked to choose two numbers for the two class sizes and determine the total number in both classes.

Creating a Sentence. Students can be asked to create a sentence that includes certain words and numbers. For example, a teacher could ask students to create a sentence that includes the numbers 3 and 4 along with the words "and" and "more," or a sentence that includes the numbers 8 and 7 as well as the words "product" and "equal." The variety of sentences students come up will often surprise teachers. In the first case, students might produce any of the sentences below and many more:

- 3 *and* 4 are *more* than 2.
- 4 is *more* than 3 *and more* than 1.
- 3 *and* 4 together are *more* than 6.
- 34 *and* 26 are *more* than 34 *and* 20.

Using "Soft" Words. Using words that are somewhat vague can allow for less precision. Rather than being negative, this can lead to a richer, more interesting conversation. For example, instead of asking for two numbers that add to 100, a teacher could ask for two numbers that add to something *close to* 100. Instead of asking students to create a triangle with an area of 20 square inches, a teacher could ask for two triangles with areas that are different but *close*.

Changing the Question. A teacher can sometimes create an open question by beginning with a question already available, such as a question from a text resource. Here are a few examples:

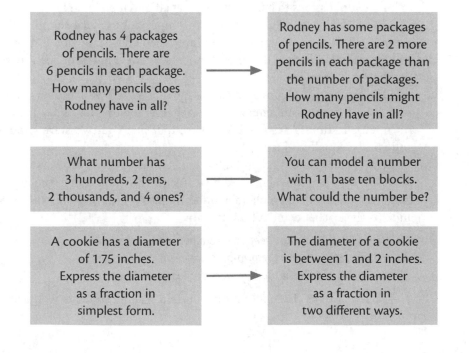

What to Avoid in an Open Question. An open question should be mathematically meaningful. There is nothing wrong with an occasional question such as *What does the number 6 make you think of?* but questions that are focused more directly on big ideas or on curricular goals are likely to accomplish more in terms of helping students progress satisfactorily in math.

Open questions need just the right amount of ambiguity. They may seem vague, and that may initially bother students, but the vagueness is critical to ensuring that the question is broad enough to meet the needs of all students.

On the other hand, one must be careful about making questions so vague that they deter thinking. Compare, for example, a question like *What's a giant step?* with a question like *How many baby steps are in a giant step?* In the first case, a student does not know whether what is desired is a definition for the term, a distance, or something else. The student will most likely be uncomfortable proceeding without further direction. In the second case, there is still ambiguity. Some students may wonder what is meant by "baby step" and "giant step," but many will be comfortable proceeding; they realize they are being asked about a measurement situation.

The reason for a little ambiguity is to allow for the differentiation that is the goal in the use of open questions. Any question that is too specific may target a narrow level of understanding and not allow students who are not at that level to engage with the question and experience success.

Fostering Effective Follow-Up Discussion. Follow-up discussions play a significant role in cementing learning and building confidence in students. Thus, it is important for teachers to employ strategies that will optimize the effectiveness of follow-up discussions to benefit students at all developmental levels.

To build success for all students, it is important to make sure that those who are more likely to have simple answers are called on first. By doing so, the teacher will increase the chances that these students' answers have not been "used up" by the time they are called on.

The teacher must convey the message that a variety of answers are appreciated. It is obvious to students when a teacher is "looking for" a particular answer. An open question is designed to ensure that many answers are good answers and will be equally valued.

The teacher should try to build connections between answers that students provide. For example, when asked how 10 and 12 are alike, one student might say that both numbers are even and another might say that they are both between 10 and 20. The teacher could follow up with:

- *What other even numbers are there between 10 and 20?*
- *Which digits tell you the numbers are even?*
- *Which digits tell you the numbers are between 10 and 20?*

Such questions challenge all students and scaffold students who need help.

Parallel Tasks

Parallel tasks are sets of tasks, usually two or three, that are designed to meet the needs of students at different developmental levels, but that get at the same big idea and are close enough in context that they can be discussed simultaneously. In other words, if a teacher asks the class a question, it is pertinent to each student, no matter which task that student completed. The use of parallel tasks is an extension of Forman's (2003) point that task modification can lead to valuable discussions about the underlying mathematics of a situation. Parallel tasks also contribute to the creation of the classroom as a learning community in which all students are able to contribute to discussion of the topic being studied (Murray & Jorgensen, 2007).

For example, suppose a teacher wishes to elicit the big idea within the NCTM Numbers and Operations strand that it is important to recognize when each mathematical operation is appropriate to use. The teacher can set out two parallel tasks:

> _Choice 1:_ Create a word problem that could be solved by
> multiplying two one-digit numbers.
>
> _Choice 2:_ Create a word problem that could be solved by
> multiplying two numbers between 10 and 100.

Both choices focus on the concept of recognizing when multiplication is appropriate, but Choice 1 is suitable for students only able to work with smaller factors. Further, the tasks fit well together because questions such as the ones listed below suit a discussion of both tasks and thus can be asked of all students in the class:

- _What numbers did you choose to multiply?_
- _How did you know how many digits the product would have?_
- _What about your problem made it a multiplying problem?_
- _What was your problem?_
- _How could you solve it?_

Strategies for Creating Parallel Tasks. To create parallel tasks to address a particular big idea, it is important to first think about how students might differ developmentally in approaching that idea. Differences might relate to what operations the students can use or what size numbers they can handle, or they might involve, for example, what meanings of an operation make sense to the students.

Once the developmental differences have been identified, the object is to develop similar enough contexts for the various choices that common questions can be asked of the students as they reflect on their work. For example, for the big idea that standard measures simplify communication, the major developmental difference might be the type of measurement with which students are comfortable. One task could focus on linear measurements and another on area measurements. One set of parallel tasks might be:

> _Choice 1:_ An object has a length of 30 cm. What might it be?
>
> _Choice 2:_ An object has an area of 30 cm². What might it be?

In this case, common follow-up questions could be:

- _Is your object really big or not so big? How did you know?_
- _Could you hold it in your hand?_
- _How do you know that your object has a measure of about 30?_
- _How would you measure to see how close to 30 it might be?_
- _How do you know that there are a lot of possible objects?_

Often, to create a set of parallel tasks, a teacher can select a task from a handy resource (e.g., a student text) and then figure out how to alter it to make it suitable for a different developmental level. Then both tasks are offered simultaneously as choices for students:

> _Original task_ (e.g., from a text):
>   There were 483 students in the school in the morning. 99 students left for a field trip. How many students are left in the school?
>
> _Parallel task:_
>   There are 71 students in 3rd grade in the school. 29 of them are in the library. How many are left in their classrooms?

Common follow-up questions could be:

- _How do you know that most of the students were left?_
- _How did you decide how many were left?_
- _Why might someone subtract to answer the question?_
- _Why might someone add to answer the question?_
- _How would your answer have changed if one more student was not in a classroom?_
- _How would your answer have changed if there had been one extra student to start with?_

Fostering Effective Follow-Up Discussion. The role of follow-up discussions of parallel tasks and the techniques for encouraging them mirror those for open questions. Once again, it is critical that the teacher demonstrate to students that he or she values the tasks equally by setting them up so that common questions suit each of them. It is important to make sure students realize that the teacher is equally interested in responses from the groups of students pursuing each of the choices. The teacher should try not to call first on students who have completed one of the tasks and then on students who have completed the other(s). Each question should be addressed to the whole group. If students choose to identify the task they selected, they may, but it is better if the teacher does not ask which task was performed as the students begin to talk.

Management Issues in Choice Situations. Some teachers are concerned that if tasks are provided at two levels, students might select the "wrong" task. It may indeed be appropriate at times to suggest which task a student might complete. This could be done by simply assigning a particular task to each student. However, it is important sometimes—even most of the time—to allow the students to choose. Choice is very empowering.

If students who struggle with a concept happen to select a task beyond their ability, they will soon realize it and try the other choice. Knowing that they have the choice of task should alleviate any frustration students might feel if they struggle initially. However, students may sometimes be able to complete a task more challenging than they first thought they could handle. This would be a very positive experience.

If students repeatedly select an easier task than they are capable of, they should simply be allowed to complete the selected task. Then, when they are done, the teacher can encourage them privately to try the other choice as well.

Putting Theory into Practice

A form such as the one that appears on the next page can serve as a convenient template for creation of customized materials to support differentiation of instruction in math. In this example, a teacher has developed a plan for differentiated instruction on the topic of measurement. A blank form is provided in Appendix B.

The following fundamental principles should be kept in mind when developing new questions and tasks:

- All open questions must allow for correct responses at a variety of levels.
- Parallel tasks need to be created with variations that allow struggling students to be successful and proficient students to be challenged.
- Questions and tasks should be constructed in such a way that all students can participate together in follow-up discussions.

Teachers may find it challenging at first to incorporate the core strategies of open questions and parallel tasks into their teaching routines. However, after trying examples found in the chapters that follow, and creating their own questions and tasks, teachers will soon find that these strategies become second nature. And the payoff for the effort will be the very positive effects of differentiation that emerge: fuller participation by all students and greater advancement in learning for all.

CREATING A MATH TALK COMMUNITY

Throughout this book, many suggestions will be offered for ways to differentiate instruction. These are all predicated on a classroom climate where mathematical conversation is the norm, a variety of student approaches are encouraged and valued, and students feel free to take risks. Unless a student engages in mathematical conversation, it is not possible for a teacher to know what that individual does or

MY OWN QUESTIONS AND TASKS

Lesson Goal: Area measurement   **Grade Level:** __3__

Standard(s) Addressed:
Geometric measurement: understand concepts of area and relate area to multiplication and to addition; recognize perimeter as an attribute of plane figures and distinguish between linear and area measures

Underlying Big Idea(s):
The same object can be described by using different measurements.

Open Question(s):
Which shape is bigger? How do you know?

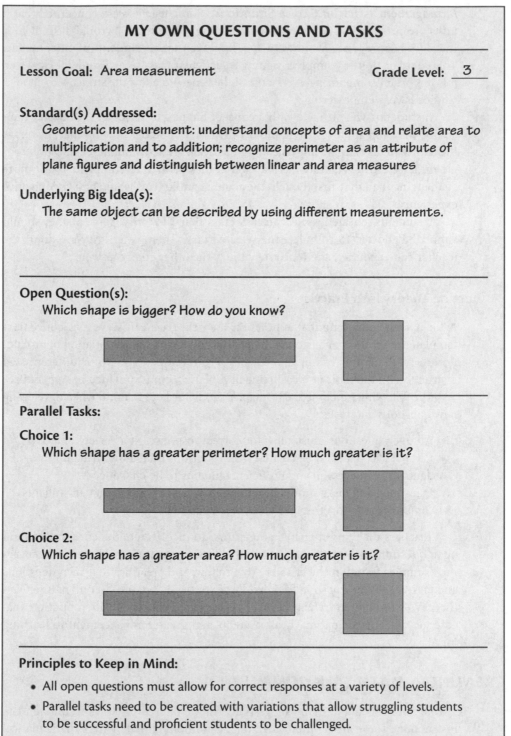

Parallel Tasks:

Choice 1:
Which shape has a greater perimeter? How much greater is it?

Choice 2:
Which shape has a greater area? How much greater is it?

Principles to Keep in Mind:

- All open questions must allow for correct responses at a variety of levels.
- Parallel tasks need to be created with variations that allow struggling students to be successful and proficient students to be challenged.
- Questions and tasks should be constructed in such a way that will allow all students to participate together in follow-up discussions.

does not understand. Consequently, it is not possible for a teacher to know what steps must be taken to ensure the student's needs are being met.

Some teachers may be nervous about offering choices or asking open questions—worried that students might get off track, worried that students might be uncomfortable with not being absolutely certain about what is expected, or worried that students might offer an idea with which the teacher is unfamiliar, leaving the teacher unsure about how to proceed. These are natural concerns.

Initially, students who are accustomed to highly structured learning environments may find open questions or choice unsettling. But once the students see the teacher's willingness to allow them to go in different directions, they grow comfortable with the change and appreciate the opportunity for greater input. Teachers also find it both surprising and rewarding to see how students rise to the challenge of engaging in mathematical conversation and how students often help the teacher sort out an unclear comment from another student or suggest ways to pick up on another student's suggestion.

Counting & Cardinality and Number & Operations in Base Ten

DIFFERENTIATED LEARNING activities in counting and cardinality and in number and operations in base ten are derived from applying the Common Core Standards for Mathematical Practice to the content goals that appear in the Counting & Cardinality domain for Kindergarten and in the Number & Operations in Base Ten domain for Kindergarten–Grade 5.

TOPICS

Before differentiating instruction in this aspect of the number strand, it is useful for a teacher to have a sense of how these topics develop over the grades.

Prekindergarten–Grade 2

Within this grade band, students move from counting and comparing very small numbers, usually 10 or less, to counting, comparing, modeling, and interpreting numbers up to 1000 using concepts that underlie the **place value system**. They move from counting to determine **sums** in simple joining situations and **differences** in simple subtraction situations to thinking more formally about adding and subtracting. Increasingly, as they move through this grade band, students use a variety of principles, strategies, and procedures with increasing efficiency to add and subtract and to solve problems requiring addition and subtraction.

Grades 3–5

Within this grade band, students begin to focus increasingly on multiplying and dividing **whole numbers** using a variety of strategies to calculate and estimate **products** and then **quotients**. They commit multiplication and related division facts to memory, become more fluent with **algorithms** for multiplying and dividing multidigit whole numbers, and solve problems that represent a variety of meanings of multiplication and division. In Grade 5, they think of decimals as part of the place value system and use place value concepts to **round** and compare both whole numbers and decimals.

THE BIG IDEAS FOR COUNTING & CARDINALITY AND FOR NUMBER & OPERATIONS IN BASE TEN

In order to differentiate instruction in the content area of number, it is important to have a sense of the bigger ideas that students need to learn. A focus on these big ideas, rather than on very tight standards, allows for better differentiation.

It is possible to structure all learning in the topics covered in this chapter around these big ideas, or essential understandings:

1.1. Counting is fundamental for describing magnitude as well as for calculating.

1.2. Numbers often tell how many or how much.

1.3. Representing a whole number in different ways tells different things about that number and might make numbers easier to compare.

1.4. The place value system standardizes how whole numbers are **decomposed**. The system makes it easier to describe, compare, count by, and calculate with numbers.

1.5. Benchmark numbers are useful for relating numbers and estimating amounts.

1.6. It is useful to take advantage of the relationships between the operations in computational situations.

1.7. Decomposing whole numbers and recomposing them are critical skills for representing, comparing, and operating with whole numbers.

The tasks set out and the questions asked about them while teaching topics in counting and cardinality and in number and operations in base ten should be developed to reinforce the big ideas listed above. The following sections present numerous examples of application of open questions and parallel tasks in development of differentiated instruction in these big ideas across the Prekindergarten–Grade 2 and Grades 3–5 grade bands.

OPEN QUESTIONS FOR PREKINDERGARTEN–GRADE 2

OPEN QUESTIONS are broad-based questions that invite meaningful responses from students at many developmental levels.

> Choose a number. Start counting. Make sure to say lots of numbers.

CCSS: Counting & Cardinality: K.CC   **BIG IDEA:** 1.1
         Number & Operations in Base Ten: 1.NBT
         Mathematical Practice: 6

Students who need to can start low, at 1 or 2, for example. But other students might want to show their counting prowess and begin with a higher number, such as 41 or even 48. All students are still practicing counting, but each is counting at his or her appropriate level.

Students who are "strong" might be encouraged to start counting at a place where transitions over decades are required, for example, counting 48, 49, and then realizing that 50 comes next.

Using the term "lots of numbers" allows for even more differentiation, with some students counting relatively few numbers and others doing more.

> Start at 0 and count by a number of your choice. What patterns do you see or hear?

CCSS: Counting & Cardinality: K.CC   **BIG IDEA:** 1.1
         Number & Operations in Base Ten: 1.NBT, 2.NBT
         Mathematical Practices: 6, 7

The reason we can count numbers we don't yet know is because of the patterns in our number sequence. Students will hear the patterns of 9 numbers ending with the words "one, two, three, . . . , nine," and will hear the transitions between decades and between centuries when counting by ones. Allowing students to skip count by numbers like 2, 5, 10, 25, 100, and so forth, alerts them to even more patterns that are possible both in the look and in the sound of the numbers they encounter.

> Samantha said it's just as easy to count by 5s as by 10s. Do you agree or disagree?

CCSS: Number & Operations in Base Ten: 2.NBT   **BIG IDEA:** 1.1
         Mathematical Practice: 3

There is, of course, no correct answer to this question. But it does give students an opportunity to put forth an argument as to why a person might think

Open Questions for Pre-K–Grade 2

that counting by 5s is as easy as or easier than counting by 10s. A student might argue that it is just as easy to count by 5s, since you use the same numbers as you would if you were counting by 10s, but you add extra numbers that end with 5s. Or students might suggest they are more accustomed to counting by 5s than 10s since they learned that earlier. The teacher might decide to have a 100 chart or a large **rekenrek** available for students to use.

TEACHING TIP. Mathematical Practice Standard 3 (constructing viable arguments and critiquing the reasoning of others) can be accessed frequently by putting out an "opinion" statement and asking students to agree or disagree.

Put out some forks and some spoons. Make sure there are just a FEW more forks than spoons.

CCSS: Counting & Cardinality: K.CC   **BIG IDEA:** 1.1
             Mathematical Practice: 6

At this level, most students are working with only very small numbers. The idea in this problem is to give students an opportunity to count, but also to get them thinking about how numbers relate to one another. The question is open in that students can put out a lot of spoons and forks if they wish, or not many. In addition, students have some latitude in deciding what "just a few" means.

It is important with this question that teachers not define the word "few" as a particular number (e.g., three), which is sometimes done. Students should be allowed to see that there is some variation in what different people consider "a few."

Arrange a group of stickers into a shape you know. Tell how many stickers you used. Tell what the shape is.

CCSS: Counting & Cardinality: K.CC   **BIG IDEA:** 1.1
             Mathematical Practice: 4

The objective in having students arrange the stickers to form a shape is partly to bring in some geometric considerations, but also to force students to use more than one or two stickers. This provides an opportunity for students to count as well as to think about shape creation.

Teachers might even make some suggestions to different students about how many stickers to use, for example, *Why don't you use more than 5 stickers?* or *Why don't you use more stickers than there are buttons on your shirt?*

> *Variations.* The use of stickers is clearly optional; students might arrange any items into a shape.

> Choose a number and start counting from there. Will you have to say a lot of numbers to get to 50 or not very many? How do you know?

CCSS: Counting & Cardinality: K.CC   **BIG IDEAS:** 1.1, 1.2
         Number & Operations in Base Ten: 1.NBT, 2.NBT
         Mathematical Practices: 2, 7

This question is made open by allowing students to choose the number they will use as a starting point. If the student chooses a low number (e.g., 2 or 3), he or she should realize it will take a while to get to 50. If the student happens to choose a number greater than 50, it will be interesting to see if she or he is comfortable counting backward as a way to reach 50. Some students will think of skip counting and, even though the numbers could be technically far apart, it might not take long to get to 50. This question should give the teacher insight into the flexibility students have with counting, while giving the students an opportunity to think about number size.

> **Variations.** The question can be varied by specifically asking students to skip count or to count backward, and the goal number 50 can be changed.

TEACHING TIP. One of the simplest strategies for differentiating instruction is allowing students to choose the numbers with which they will work.

> A number describes how many students are in a school. What might the number be?

CCSS: Number & Operations in Base Ten: 1.NBT, 2.NBT   **BIG IDEA:** 1.2
         Mathematical Practices: 2, 3, 4

An important part of mathematics instruction is to allow students to make sense of the numbers around them. People need to be able to recognize when a number makes sense and when it does not. For example, students should know that a price of $500 for a pencil makes no sense.

The question above illustrates a real-life context. In this case, the situation is close to students' everyday experience; they can think about how many students are in their own school. However, the question can also lead them to consider other possibilities, for example, how many students there might be in a big high school, a small country school, or a medium-sized school.

Some students will use calculations, adding the number of students in each class. If students simply guess, they should be expected to justify their values.

> Choose a number between 1 and 100. Tell why someone might call it a big number, but someone else might call it a small one.

CCSS: Counting & Cardinality: K.CC   **BIG IDEA:** 1.2
        Number & Operations in Base Ten: 1.NBT, 2.NBT
        Mathematical Practice: 3

An important mathematical concept for students to learn is that no number is inherently big or little; it always depends on context. The number 10 could be big if it describes the number of children in a family, but it would be small if it tells how many customers are in a large store.

➤ *Variations.* This kind of question is one that could be repeated over and over using numbers of the students' choice or the teacher's choice.

> What makes 5 a special number?

CCSS: Counting & Cardinality: K.CC   **BIG IDEAS:** 1.2, 1.3
        Mathematical Practices: 2, 3

Posing this question to young students provides them an opportunity to participate in a mathematical conversation. Some students might think of the fact that there are 5 fingers on a hand, others that they are 5 years old, others that there are 5 people in their family, whereas others might think of something else, for example, that there is a special coin for 5 cents. The question helps students recognize that numbers are used to describe amounts in a wide variety of contexts.

➤ *Variations.* This question can be varied by using other numbers that students might view as special, for example, 0, 1, or 10.

> Choose two of these numerals. How do they look the same? How do they look different?
>
> 1   2   3   4   5   6   7   8   9   0

CCSS: Counting & Cardinality: K.CC   **BIG IDEA:** 1.3
        Mathematical Practice: 3

An important part of mathematical development is the recognition that numbers can be used to represent quantities, but it helps if students recognize the numerals. By focusing on the form of the numerals, this question helps students learn to reproduce and read them.

By allowing students to choose whatever two numerals they wish, the question allows for virtually every student to succeed. For example, a very basic comparison might be that 1 and 4 both use only straight lines, but fewer lines are needed to form a 1 than to form a 4.

> Show the number 7 in as many different ways as you can.

CCSS: Counting & Cardinality: K.CC   **BIG IDEA:** 1.3
        Mathematical Practice: 5

To respond to this question, some students will draw seven similar items, others will draw seven random items, others will use stylized numerals, others will write the word *seven,* and still others will show mathematical operations that produce an answer of 7, such as 6 + 1 or 4 + 3.

Because the question asks students to show as many representations as they can, even a student who comes up with only one or two representations will experience success. No matter what their developmental levels, all students will focus on the fact that any number can be represented in many ways. It is important to pose questions that focus students on how their own representations are alike and different.

For example, a teacher could ask:

* *Which of your representations are most alike?*
* *Which of your representations make it easy to see that 7 is less than 10?*
* *Which make it easy to see that 7 is an odd number?*
* *Which make it easy to see that 7 is 5 and 2?*

> The answer is 5. What is the question?

CCSS: Counting & Cardinality: K.CC   **BIG IDEA:** 1.3
        Number & Operations in Base Ten: 1.NBT
        Mathematical Practices: 3, 5

This question has innumerable possible answers, and it supports the important concept that numbers can be represented in many ways. Some students will use addition and ask the question: *What is 4 + 1?* Other students will use subtraction and ask: *What is 6 − 1?* Yet other students will ask very different types of questions, for example: *What number comes after 4?* or even *How many toes are there on one foot?* Because of the wide range of acceptable responses, students at all levels are addressed.

> **Variations.** This task can be reassigned after changing the number required for the answer.

> You can represent a number between 100 and 1,000 with just a few **base ten blocks**. What could the number be? How do you know?

CCSS: Number & Operations in Base Ten: 2.NBT   **BIG IDEAS:** 1.3, 1.4, 1.7
        Mathematical Practices: 2, 7

Using base ten blocks focuses students on the nature of the place value system. Students consider what "columns" there are in a **place value chart**, because each block type represents a different column, and they also consider relative size, for example, that the 100 is ten times as great as the 10.

There will be students who will realize that a relatively large number (e.g., 200) can sometimes be represented by many fewer blocks than a much smaller number (e.g., 17). Others will simply randomly select one, two, or three blocks from a set of blocks; their focus will then be on naming the number.

It would be interesting to observe whether students first choose blocks and then name their numbers or whether they realize that they want a number with zeroes in it with the other digits fairly small.

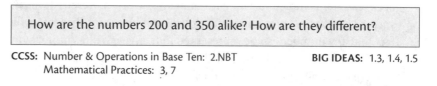

How are the numbers 200 and 350 alike? How are they different?

CCSS: Number & Operations in Base Ten: 2.NBT   **BIG IDEAS:** 1.3, 1.4, 1.5
Mathematical Practices: 3, 7

A question like this one provides the opportunity to see what students know about numbers in the hundreds. They might observe similarities such as these:

- *They are both more than 100.*
- *They are both less than 400.*
- *You say both of them when you skip count by 10.*
- *They both have three digits.*
- *If they were an amount of cents, you could model both of them with quarters.*

200:

350:

The students might observe differences such as these:

- *350 is more than 300, but 200 is not.*
- *200 is an exact number of hundreds, but 350 is not.*
- *You can count from 0 to 200 by 20s, but you can't count by 20s to 350 if you start at 0.*

The question is suitable for a broad range of students because some students can choose simple similarities and differences, such as indicating that both numbers end in 0, whereas others can focus on more complex similarities, such as indicating that both numbers are part of the sequence when skip counting by 25. All students benefit from the larger discussion about number representations and meanings.

> ➤ ***Variations.*** The question can easily be varied by using other pairs of numbers.

> Two numbers each have 7 as one of their digits. But one 7 is worth a LOT more than the other. What could the numbers be?

CCSS: Number & Operations in Base Ten: 1.NBT, 2.NBT   **BIG IDEA:** 1.4
Mathematical Practices: 1, 2, 3, 7

To really understand the place value system, a student needs to realize that the placement of a digit matters a lot. If, for example, a 7 is in the ones place, it is clearly worth less than if it is in the tens or hundreds place. But the other digits are irrelevant. So students might choose, for example, 7 and 700, or 17 and 79, or 27 and 712.

Some students might benefit from access to a place value chart or base ten blocks.

> A two-digit number has more tens than ones. What could the number be? How do you know your number is correct?

CCSS: Number & Operations in Base Ten: 1.NBT, 2.NBT   **BIG IDEAS:** 1.4, 1.5
Mathematical Practices: 2, 3, 7

Much of the work with numbers is built on the fact that numbers are written in such a way that the value of a numeral is dependent on its placement in the number. For example, the 2 in 23 is worth 20, but the 2 in 32 is worth only 2.

To help students work with the place value system, instruction often begins with models that show the difference. For example, base ten blocks show 20 as two rods, but 2 as two small cubes.

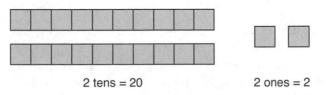

2 tens = 20   2 ones = 2

Students may or may not be provided with base ten blocks when the question above is posed. For most students, provision of the blocks would be helpful, but if students are more advanced, the blocks may not be essential. Some students may discover only one or two possible responses, whereas others might determine many possible responses (e.g., 31, 32, 43, 42, 41, etc.) or even all possible responses. By asking for only one number, the task seems less onerous to struggling students. In discussing responses, students will see that there were many possible values among which they could have chosen.

In showing how they know their number is correct, some students can use concrete models to support their answers, whereas others might use more symbolic arguments.

> *Variations.* Students who are ready to work with three-digit numbers can be given the option to do that if they wish.

One number is a lot more than another one. What could the two numbers be?

CCSS: Number & Operations in Base Ten: 1.NBT, 2.NBT   **BIG IDEAS:** 1.4, 1.5
        Mathematical Practices: 2, 7

This question provides a great deal of latitude both in the choice of numbers and in the interpretation of the phrase *a lot more.*

For example, one student might choose 200 and 300, suggesting that being 100 apart means that 300 is a lot more. Another student might choose 10 and 90. Others might argue that those numbers are not far enough apart and might choose, for example, 1,000,000 and 101. None of these students are incorrect, but the door has been opened for a discussion about how some mathematical terms are more precise than other terms and how both types of terms can be useful in different situations.

In addition, there is an opportunity within the larger class community to repeatedly practice the concept of comparison.

Think of a number made up of tens and ones. Switch the number of tens and the number of ones. What happens to the value of your number? Why?

CCSS: Number & Operations in Base Ten: 1.NBT, 2.NBT   **BIG IDEAS:** 1.4, 1.5
        Mathematical Practices: 2, 3, 7

Much like an earlier question, this question, in a slightly different way, focuses on the difference in the value of a digit based on its placement in a number. Some students might select numbers where the value decreases; they might note that this occurs when the tens digit is the greater digit in the original number (e.g., 43, 75, 91). Others might select numbers where the value increases; they might note that this occurs when the tens digit is less than the ones digit in the original number (e.g., 34, 57, 19). It will be interesting if some students pick a number with a repeated digit (e.g., 22) or a number where the ones digit is 0 (e.g., 30). In the latter case, they will have to discuss what 03 actually means.

> *Variations.* A similar question could be used with greater numbers with different digits being exchanged.

> Choose three numbers. Two have to be pretty close together. The third has to be far from the other two. What might they be?

CCSS: Counting & Cardinality: K.CC   **BIG IDEAS:** 1.4, 1.5
             Number & Operations in Base Ten: 1.NBT, 2.NBT
             Mathematical Practices: 2, 3

This question provides insight into what students think close together and far apart mean for numbers. It also gives them an opportunity to use numbers that are comfortable for them, making the question accessible to most students. While some students might choose numbers like 2, 3, and 9, others might choose numbers like 95, 100, and 2, or 500, 510, and 900. By asking students to write the numbers, it is more likely they will choose numbers that are not too large.

It would be interesting to discuss and compare answers. A teacher might ask:

- *Do you think that 100 and 110 are close?*
- *Do you think that 5 and 8 are close?*
- *How did you decide which numbers to choose?*

> Use a **100 chart**. Choose two numbers to add. Both numbers must be on the top half of the chart. Show how to use the chart to add the two numbers without using pencil or paper.

1	2	3	4	5	6	7	8	9	10
11	12	13	14	15	16	17	18	19	20
21	22	23	24	25	26	27	28	29	30
31	32	33	34	35	36	37	38	39	40
41	42	43	44	45	46	47	48	49	50
51	52	53	54	55	56	57	58	59	60
61	62	63	64	65	66	67	68	69	70
71	72	73	74	75	76	77	78	79	80
81	82	83	84	85	86	87	88	89	90
91	92	93	94	95	96	97	98	99	100

CCSS: Number & Operations in Base Ten: 1.NBT, 2.NBT   **BIG IDEAS:** 1.4, 1.7
        Mathematical Practices: 3, 5

This question promotes students' use of visualization to help them with **mental math**. They can select simple numbers to add or more complicated ones. For example, students who start at 32 and add 1 realize that it is only necessary to move one space to the right. If the student adds 10, it is only necessary to move one space down. A student who adds something like 29, though, might realize that it is possible to go down three rows and back one space, a much more sophisticated mental math operation.

The question is useful for a variety of student levels because students can avoid adding to a number at the right side of the chart (e.g., adding 59 + 6) and having to go to the next line and can avoid adding complicated numbers if they find the maneuvers too difficult.

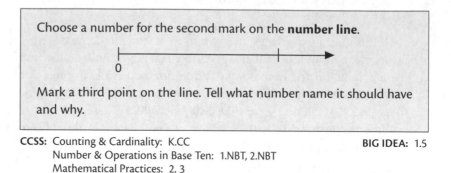

Choose a number for the second mark on the **number line**.

Mark a third point on the line. Tell what number name it should have and why.

CCSS: Counting & Cardinality: K.CC   BIG IDEA: 1.5
Number & Operations in Base Ten: 1.NBT, 2.NBT
Mathematical Practices: 2, 3

This question helps students see the value of using benchmarks to relate numbers. It is designed to be open by allowing students not only to choose which point to locate but also which benchmark to use.

A simple response might be to designate the second marked point as a 2 and to mark a halfway point as a 1. A more complex response might be to designate the second marked point as a 10 and attempt to locate a number like 2 or 3.

No matter which choice students make, they can participate in the conversation about how numbers are placed on a number line and how numbers relate to one another.

➤ *Variations.* The teacher can name the second point, for example as 5, then ask students to choose the third point.

A news item says that about 50 people were at a certain event.
Exactly how many people might that have been?

CCSS: Number & Operations in Base Ten: 1.NBT   BIG IDEA: 1.5
Mathematical Practices: 2, 3

Notice that there is no indication of how the number 50 was obtained. Perhaps a count was rounded to the nearest 10 or to the nearest 25, or maybe it was not rounded at all. It is quite possible that the original number could have been something like 38, since 38 is not that far from 50. Or maybe it was greater, perhaps 59 or 60. Because there is no indication of how the estimate was obtained, this question is open.

➤ *Variations.* The number 50 could be changed.

Open Questions for Pre-K–Grade 2

> The sum of ☐☐ and ☐☐ is closer to 80 than to 90. What could the two numbers be?

CCSS: Number & Operations in Base Ten: 1.NBT, 2.NBT   **BIG IDEAS:** 1.5, 1.7
     Mathematical Practices: 1, 2, 5, 7

It would be interesting to see how students approach this question. Most likely, students will try to think of numbers that sum to 81, 82, 79, or some other number close to 80. But some students will realize that any sum less than 80 (even one as small as 20) is closer to 80 than to 90, and so they might use numbers such as 10 and 10. Some may push the limits to 84 for their sum.

It would be valuable to have a discussion about how students broke up the numbers 80 and/or 90 to help them answer the question.

➤ *Variations.* Instead of asking for a sum closer to 80 than to 90, values other than 80 and/or 90 could be selected.

> Describe a subtraction calculation that would be easy to do if you realized how addition and subtraction are related. Why would it be easy?

CCSS: Counting & Cardinality: K.CC   **BIG IDEA:** 1.6
     Number & Operations in Base Ten: 1.NBT, 2.NBT
     Mathematical Practices: 2, 3, 5, 7

Students might choose smaller numbers such as 10 – 6 and think that since they know that 6 + 4 = 10, it is easy to figure out 10 – 6. Or a student might use a 100 chart and point to a number on the chart and realize it is easy to figure out 100 minus that number by counting how many more spaces there are on the chart. Or a student might decide that 99 – 38 is easy to solve by figuring that she or he needs to add 1 one and 6 tens (or 61) to get to 99.

TEACHING TIP. It is always safer to ask students for something that they find easy than for something they find difficult. There is no risk involved in indicating something is easy for you.

> You add two numbers that are almost 30 apart. The answer is almost 90. What might the numbers be?

CCSS: Number & Operations in Base Ten: 1.NBT, 2.NBT   **BIG IDEAS:** 1.6, 1.7
     Mathematical Practices: 1, 2, 5, 7

A number of different strategies might be employed to solve this problem. Most students will use guess and check, but in different ways.

Whereas some students will try a combination, for example, 10 + 40, realize that the answer is too small, and just move up to 20 + 50 and then 30 + 60, others will realize right away, with 10 + 40, that they are 40 short and that each number has to go up by 20.

Other students might begin with a combination for 90, for example, 40 + 50, realize that the two **addends** are only 10 apart, and move the 40 down and the 50 up to get the appropriate difference. Both of these approaches involve valuable reasoning.

The fact that the question says "almost" means that students will probably take those answers and figure out how to adjust them. For example, the student might use 30 + 59, so that the numbers are not quite 30 apart and the answer is not quite 90.

> *Variations.* There are many ways to change this question. The numbers used could be changed or the word *almost* might be changed to *about* or *a bit more than*.

Make up an addition question where there is a 2, a 3, and a 4 somewhere in the question or the answer.

CCSS: Number & Operations in Base Ten: 1.NBT, 2.NBT   **BIG IDEA:** 1.7
Mathematical Practices: 1, 2, 3, 5

All students who respond to the question will use addition, but different students can choose combinations with which they are more comfortable. For example, students might write:

$$\underline{2} + \underline{34} = \underline{36} \qquad \underline{2} + \underline{3} + \underline{4} = 9 \qquad \underline{23} + \underline{43} + \underline{25} = 91$$

An open question such as this promotes a rich discussion. As different individuals or pairs of students share their questions, other students repeatedly have the opportunity to learn new ideas about addition.

> *Variations.* The question can easily be varied by using different sets of three numbers, using two or four numbers, or using a different operation (e.g., subtraction).

How is adding 42 + 38 like adding 52 + 28? How is it different?

CCSS: Number & Operations in Base Ten: 1.NBT, 2.NBT   **BIG IDEA:** 1.7
Mathematical Practices: 2, 3, 7

This question is interesting in that it allows students to focus on a variety of things. Some will focus on a numerical process, others on models, and still others on the meaning of addition. Possible answers might be:

- *They are alike since both are adding, but they use different numbers.*
- *They are alike since both involve adding 2 + 8, but different since one involves 40 + 30 and the other involves 50 + 20.*
- *They are alike since they have the same answer, but they use different numbers.*

- *They are alike since if you use **10-frames** you would end up with full frames both times even though you started with a number where there was a partial frame. The actual number of frames you start with is different.*

Ultimately, it would be helpful for students to see that when 10 is moved from the second addend to the first, the total does not change.

OPEN QUESTIONS FOR GRADES 3–5

> Describe 100 thousand in as many ways as you can.

CCSS: Number & Operations in Base Ten: 4.NBT   **BIG IDEAS:** 1.2, 1.3, 1.4, 1.5
Mathematical Practices: 2, 7

Typically this question would be presented using the numeral 100,000, rather than the written phrase *100 thousand*. The advantage of the suggested presentation is that the question becomes more accessible to students who struggle with numerals that represent large numbers. Although one might think that students need to have a concept of how much 100,000 is to answer the question, which would certainly be preferable, a student could simply say that it is more than 99 thousand.

Other students might refer to 100,000 by using place value ideas, for example, indicating that it is possible to represent 100,000 with 100 large base ten blocks, that it is $\frac{1}{10}$ of 1 million, or that it can be represented with 1,000 hundred blocks.

The class discussion would provide ample opportunity for exploration as different aspects of 100 thousand are raised.

> How are the numbers 6.001 and 1.006 alike? How are they different?

CCSS: Number & Operations in Base Ten: 5.NBT   **BIG IDEA:** 1.4
Mathematical Practice: 3

Responses to this open question might involve how the numbers can be represented on a place value chart or with base ten blocks, their relative sizes, or simply their digits.

Whereas some students will concentrate on the fact that both numbers use the same digits (thus classifying the numbers in terms of the digits that make them up), others will recognize that one is greater than 6 and one less than 6 (thus classifying the numbers in terms of their relationship to 6), that both involve decimal thousandths (thus classifying the numbers in terms of their decimal representations), or that both are less than 10. Even a struggling student will recognize that the same digits are involved.

➤ *Variations.* In the example cited, decimal numbers were compared and contrasted. New questions can be created in which pairs of fractions, **mixed numbers**, or large numbers such as 1.1 million and 1.1 billion are compared.

In a particular multidigit number, there are two 7s and two 5s. One 7 is worth 100 times as much as the other; one 5 is worth 10 times as much as the other. What could the number be?

CCSS: Number & Operations in Base Ten: 5.NBT   **BIG IDEA:** 1.4
       Mathematical Practice: 7

When students really understand the place value system, they realize that a digit one column to the left of the same digit is worth 10 times as much as the one on the right. Similarly, a digit two columns to the left is worth 100 times as much as the one on the right. This question builds on that knowledge. There are many correct answers, for example, 71,755 or 55,727 or 55,797. But some students might use even larger numbers, such as 707,554 or 1,717,455; or others might use decimals, perhaps something like 55.727.

You multiply two numbers and the product is of the form ☐2,☐4☐. What could the numbers and product have been?

CCSS: Number & Operations in Base Ten: 5.NBT   **BIG IDEA:** 1.4
       Mathematical Practices: 1, 6

There are many solutions to this problem. Leaving the ones digit unknown might seem, at first glance, to make the question more difficult; often we use the ones digit to figure out what the ones digits of the **factors** might have been. In fact, though, leaving the ones digit open makes the problem more accessible. For example, putting a 0 in the ones digit allows students to use factors like 10×3254 or 20×3142.

Some students might simply choose a 2-digit number at random and multiply by other numbers until they hit the desired form. They will soon realize that in order to get a 5-digit answer, they are likely to be multiplying by a 3-digit or 4-digit factor; this helps build their number sense.

Other ideas that might come up that are worth discussing include looking at whether factors and the product are even or odd, whether they are **multiples** of particular numbers (such as 5), and so forth.

You multiply two numbers and the product is almost 400. What could the numbers have been? Explain your answer.

CCSS: Number & Operations in Base Ten: 4.NBT   **BIG IDEAS:** 1.4, 1.7
       Mathematical Practices: 2, 3, 7

Estimation is an important aspect of mathematical calculation. Teachers often do not emphasize it enough. This open question allows students to work backward and think about what numbers might have been used to arrive at a particular estimate. They will also have to think about what "almost" means.

Some students might think of something fairly simple, for example, 1 and 399. Others might recognize that $20 \times 20 = 400$ and then use 19×19. Still others will consider other possibilities.

> **Variations.** Instead of using "almost," the question can be varied by using "a bit more than." The product can be changed from 400 to a different number. Other operations can also be allowed: for example, the question might be worded, "You add, subtract, multiply, or divide two numbers and the result is about 400. What could the numbers be? Explain."

You are going to add a 2-digit or a 3-digit number to 348. You do not find it too difficult to do the addition in your head. What might you be adding? Why is it fairly easy to do in your head?

CCSS: Number & Operations in Base Ten: 3.NBT   **BIG IDEAS:** 1.4, 1.7
Mathematical Practices: 2, 3, 7

This question allows students to use a variety of strategies, most likely based on place value, to perform mental calculations. The student might add multiples of 10 or 100, such as 10, 20, 100, 200, or might even decide to add a number such as 99, recognizing it is fairly easy to add 100 and subtract 1.

When asked a question like this, students are hearing the important message that adding mentally is a valuable skill, but which calculation they actually do can vary depending on their readiness to deal with different sorts of calculations.

> **Variations.** Additional questions could be created by changing the operation or changing the sum.

A number is about 400. What is the least you think it can be? What is the greatest?

CCSS: Number & Operations in Base Ten: 3.NBT   **BIG IDEA:** 1.5
Mathematical Practices: 2, 3

Although we often ask students to estimate or round numbers, we less frequently ask them to think about what a number might have been that has already been rounded. If you read a newspaper headline that says "About 400 people attended the meeting," you want to know what the actual number might have been.

Notice that there was no indication here whether the 400 was the result of rounding to the nearest 10 or the nearest 100 or whether rounding was even used, and that is what makes the question interesting. Different students will interpret it in different ways.

Questions a teacher might ask could include:

- *Do you think it could have been 398?*
- *Could it have been more than 400? How much more?*
- *Could it have been as low as 325? 350?*

> Choose a 4-digit factor and a 2-digit factor. Show how you could represent the product with models or diagrams in a way that would help you figure out the answer.

CCSS: Number & Operations in Base Ten: 5.NBT
Mathematical Practice: 5

BIG IDEA: 1.5

By giving students a choice of values, they can make the situation simpler (e.g., 1000×20) or more complicated (e.g., 3142×5).

To represent 1000×20, a student might use a place value chart and place two counters in the tens column to show 20; multiplying by 1000 has the effect of moving the counters three columns to the left.

Ten Thousands	Thousands	Hundreds	Tens	Ones
			⬤ ⬤	

⬇

Ten Thousands	Thousands	Hundreds	Tens	Ones
⬤ ⬤				

To show 3142×5, a student might draw a table like this one:

×	3	1	4	2
5	15	5	20	10
Regrouped	15	7	1	0

> You have a decimal thousandth number. When you round it to the nearest tenth, you round down. But when you round it to the nearest hundredth, you round up. What might the decimal thousandth number be?

CCSS: Number & Operations in Base Ten: 5.NBT
Mathematical Practices: 1, 7

BIG IDEA: 1.5

Students learn the rules of rounding and may apply them, but often they don't actually think about them. This question forces students to think about the rules. Students will see that the only way it is possible to round down to the nearest tenth but up to the nearest hundredth is to have a decimal of the form ☐.*abc,* where *b* is less than 5, but *c* is 5 or more. For example, 13.928 rounds to 13.9 (less than 13.928) or 13.93 (more than 13.928).

Students should be led to see that the values of the whole number and the tenths part of the decimal are irrelevant to this problem.

> Choose a subtraction problem you want to complete where the answer is in the hundreds. Show how you could figure out the answer to the problem efficiently either by adding or by subtracting.

CCSS: Number & Operations in Base Ten: 5.NBT   **BIG IDEA:** 1.6
Mathematical Practices: 5, 6

A student might choose a calculation like 500 – 198. The student might then either add up from 198 to get to 500 or, for example, subtract 498 – 198 and then add 2.

Allowing students to choose the values they will work with empowers them and likely leads to greater success for more individuals.

> You divide two whole numbers and the quotient has 2 digits with no **remainder**. How many digits might the **dividend** and **divisor** have?

CCSS: Number & Operations in Base Ten: 5.NBT   **BIG IDEAS:** 1.6, 1.7
Mathematical Practices: 6, 7

An important idea for students to learn is that with division, as with subtraction, you can end up with a relatively small answer when using either small numbers or large numbers. With subtraction, the result is 2 whether you subtract 2 from 4 or you subtract 30,002 from 30,004. Similarly, with division, the result is 2 whether you divide 4 by 2 or you divide 40,000 by 20,000. This open question addresses that idea in that knowing the answer has 2 digits does not require the use of large numbers, but it does allow for it.

Another concept that is addressed in this question is the relationship between multiplication and division. Since we know there is no remainder, we know that we could multiply the divisor by a 2-digit quotient to achieve the dividend. Therefore, to come up with a solution, a student might think $5 \times 23 = 115$, so the question could have been $115 \div 5$. Similarly, $334 \times 10 = 3340$, so the question could have been $3340 \div 334$. Notice that in the first case, the divisor and dividend had 1 and 3 digits, respectively, and in the second case 3 and 4 digits, respectively. Students should note that the difference between the number of digits always seems to be 1 or 2 and then might investigate why that is the case.

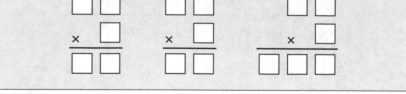

CCSS: Number & Operations in Base Ten: 4.NBT   BIG IDEA: 1.7
      Mathematical Practices: 1, 2, 7

This question requires students to multiply two-digit numbers by single-digit numbers but gives them a choice about what numbers to use.

Students who have more difficulty with multiplying may choose to multiply only by 1 and 2 to make the multiplication easier (e.g., 34 × 1 = 34, 42 × 1 = 42, 62 × 2 = 124). Other students might try other combinations, having to take into account that the products must involve only certain digits. For example, a student might begin by multiplying 4 by 2 and realize that this is not allowed because the product is 8, a digit that cannot be used in this case.

All students will have the chance to practice multiplication, using whatever strategies they prefer.

TEACHING TIP. If a mathematically strong student always seems to select values that make a question too easy, the teacher should allow the student to make the initial choice, but then challenge him or her to try other values as well.

> 4 is a factor of two different numbers. What else might be true about both of the numbers?

CCSS: Number & Operations in Base Ten: 4.NBT   BIG IDEA: 1.7
      Mathematical Practices: 3, 7

A critical aspect of mathematical thinking is the recognition that knowing one thing about a number can provide other information about that number. For example, knowing that a number is greater than 100 automatically means it is also greater than 10.

By knowing that a number is a multiple of 4 (because 4 is its factor), a student should be able to tell that the number is even and that it ends in 0, 2, 4, 6, or 8. Some students might even realize that if one subtracts 4 from the number, or adds 4 to the number, the result is also a multiple of 4. It is up to the teacher, by prompting other students, if necessary, to elicit the idea that this is still a **conjecture**.

There may be students who can explain why the sum or difference must be a multiple of 4.

With the question posed in the way indicated, a student who is less likely to generalize can still select two specific numbers and tell something true about both of them. For example, the student might choose 8 and 12 and point out that both are even.

➤ *Variations.* Similar questions can be posed by changing the number that is the factor.

> You add two decimals and then multiply by 5. The result is between 4.5 and 5.0. What could the decimals have been?

CCSS: Number & Operations in Base Ten: 5.NBT   BIG IDEA: 1.7
       Mathematical Practices: 1, 6

Because two operations are performed, there is significant latitude in coming up with an answer here. Some students might just do a guess-and-test and then revise their answers. For example, they might start with 1.2 + 3.4 to get 4.6. They quickly see that multiplying their sum by 5 will get take them beyond 5.0, so they reconsider. One possibility is to just start all over again; that's what most students will do. A more sophisticated response is to notice that their original sum is within the desired range. So, if they would divide their two original decimals by 5 and then multiply the new sum by 5 (as the problem requires), they would be in the right range. They readjust their original numbers to $1.2 \div 5 = 0.24$ and $3.4 \div 5$, which is about 0.7. Then, adding $0.24 + 0.7 = 0.94$. And $0.94 \times 5 = 4.7$, a suitable result.

Other students might work backward. They might say, *I want an answer of 4.8 after I multiply by 5, so I want an answer of $4.8 \div 5 = 0.96$ before I multiply.* Then they choose two decimals that add to 0.96. There are, of course, many possibilities.

Still other students might decide that they want to make the second decimal super small, so really they just need to find a first number that they can multiply by 5 and get between 4.5 and 5. Then when they add their tiny decimal to it, and multiply by 5, they will still be in the right range. They realize that 0.9 might work, so they choose 0.9 and 0.001 as their decimals to add.

> You multiply two numbers, each greater than 100. The hundreds digit of the product is 4. What could the numbers be?

CCSS: Number & Operations in Base Ten: 5.NBT   BIG IDEA: 1.7
       Mathematical Practices: 1, 6, 7

Some students will analyze the situation and realize that the hundreds digit of the product of *abc* and *efg* comes from $ag + bf + ce$ and possibly adding in some carryover from the tens digit.

If there is no carryover, a student might arbitrarily choose values for b and f to determine appropriate values for a, c, e, and g. For example, values of $b = 0$ and $f = 0$ would result in the need for $ag + ce$ to end in 4. That could happen if $a = 2$, $g = 4$, $c = 3$, and $e = 2$. The original numbers, therefore, could be 203 × 204.

Another student might look for $ag + bf + ce$ to end in, for example, 2, but with a carryover of 2 from the tens place. The tens digit comes from calculating $bg + cf$ and recording the final digit (assuming no carryover from the ones place). To make this result in 20 or more (so there is a regroup of 2 tens), values of $b = 3$, $g = 4$, $c = 4$, and $f = 2$ could work. Then the value $ag + bf + ce$ has to end in 2, but based on values for b, g, c, and f previously selected, it is the value $4a + 6 + 4e$ that has to end in 2. That could happen if $a = 3$ and $e = 1$, so the two numbers might be 334 and 124. The product works; it is 41,416.

Still other students will guess and test and get in a lot of computational practice.

> **Variations.** The value of the hundreds digit need not be 4; any value could be chosen. Or it could be the tens digit that is required to be a particular value, and not the hundreds digit.

PARALLEL TASKS FOR PREKINDERGARTEN–GRADE 2

PARALLEL TASKS are sets of two or more related tasks that explore the same big idea but are designed to suit the needs of students at different developmental levels. The tasks are similar enough in context that all students can participate fully in a single follow-up discussion.

> _Choice 1:_ Count by tens from 10 to 100. How many numbers do
>   you say?
>
> _Choice 2:_ Count by tens from 10 to 30. How many numbers do
>   you say?

CCSS: Counting & Cardinality: K.CC   **BIG IDEA:** 1.1
        Number & Operations in Base Ten: 1.NBT
        Mathematical Practice: 7

Some students will be ready to count by tens earlier than others. This parallel task addresses that reality.

Questions a teacher could ask both groups of students include:

• _How far apart are your numbers on a number line?_
• _Do you get to high numbers quickly or slowly?_
• _What pattern do you notice in the numbers you say?_

> *Choice 1:* Draw a lot of dots so that it is fairly easy to tell how many there are. How many dots did you draw, and why is it easy to tell how many there are?
>
> *Choice 2:* Draw not too many dots, but more than 3. Tell how many there are. How could you arrange them so it is easy to tell how many there are?

CCSS: Counting & Cardinality: K.CC
Mathematical Practice: 3

BIG IDEA: 1.1

Subitizing is a critical step in developing an understanding of number in young students. This sort of question focuses on subitizing but allows for both more and less sophisticated approaches. For example, some students will simply arrange 5 dots as they appear on a die and recognize the 5. Others might arrange 20 dots in four groups of 5 dots, each group as they appear on a die, skip counting to say "5, 10, 15, 20."

Questions a teacher could ask both groups of students include:

- *Why didn't you just arrange the dots all over the page?*
- *What makes it easy to tell how many dots there are?*
- *Is there a different way to arrange the same number of dots to make it easy to tell how many there are?*

> *Choice 1:* Use counters to make it easy to see whether 10 or 7 is more and how much more.
>
> *Choice 2:* Use counters to make it easy to see whether 5 or 2 is more and how much more.

CCSS: Counting & Cardinality: K.CC
Mathematical Practices: 4, 5

BIG IDEA: 1.1

These two tasks are similar in that both provide opportunities for students to compare numbers and both require that students think about how to arrange quantities to make it easy to compare them. The first choice uses slightly higher numbers than the second.

Whichever task students perform, a teacher might ask questions like these:

- *How did you know that your big number was not 1 more than your little one?*
- *If you made just two piles, one for each number, would it be easy to tell how much more was in the big pile?*
- *How did you arrange your counters to make it easy to see how much more one of the numbers was than the other?*

Parallel Tasks for Pre-K–Grade 2

> **Choice 1:** Choose a number that could tell how many flowers might fit in a vase. Tell why that number makes sense.
>
> **Choice 2:** Choose a number that could tell the number of families in a small town. Tell why that number makes sense.

CCSS: Counting & Cardinality: K.CC   **BIG IDEA:** 1.2
      Number & Operations in Base Ten: 1.NBT, 2.NBT
      Mathematical Practices: 2, 3

The number of flowers that fits in a vase is likely to be a fairly small number. However, there is some flexibility. A student might think of a larger vase that could hold 20 or more flowers, but the number is unlikely to be in the hundreds.

The number of families in a small town is likely to be a two-digit, three-digit, or maybe even a four-digit number, depending on the individual student's personal view of what constitutes a small town. The discussion would be an interesting one as students share their personal views on what numbers are reasonable and why.

All students could be asked how they decided on their numbers and why they thought their numbers were reasonable.

> **Choice 1:** Choose a number between 1 and 10. Show that number in as many ways as you can.
>
> **Choice 2:** Choose a number between 20 and 30. Show that number in as many ways as you can.

CCSS: Counting & Cardinality: K.CC   **BIG IDEA:** 1.3
      Number & Operations in Base Ten: 1.NBT
      Mathematical Practice: 5

The difference between the two versions is very slight, but providing a choice allows students who are working at different mathematical levels to select the more appropriate task.

Students might represent their numbers by using pictures, numbers, or words. For example, a student might represent 9 as $10 - 1$, as a picture of 9 items, as 9 dots in a 10-frame, as $4 + 5$, as 3 groups of 3, or by using stylized versions of the numeral 9.

No matter which task was chosen, students could be asked:

- *What number did you represent?*
- *How do you know that that number is one that was okay to choose?*
- *What are some of the different ways you represented that number?*

TEACHING TIP. Changing the values students work with, but using the same question, is often an easy way to create parallel tasks.

> ***Choice 1:*** Choose two numbers between 10 and 20. Represent each with 10-frames. How are the models alike and different?
>
> ***Choice 2:*** Choose two numbers between 44 and 55. Represent each with 10-frames. How are the models alike and different?

CCS: Counting & Cardinality: K.CC   **BIG IDEA:** 1.3
Number & Operations in Base Ten: 1.NBT
Mathematical Practice: 3

The two choices differ only in terms of the size of number with which students might be comfortable. Both involve using simple 10-frame models. Although the numbers in ***Choice 1*** will all result in one full frame and one partial frame, the numbers in ***Choice 2*** might involve four or five full frames.

Questions that suit both groups of students include:

- *What were your two numbers? How did you choose them?*
- *How did you know they were the right size?*
- *Why did you need more than one 10-frame?*
- *Why didn't you need ten 10-frames?*
- *How were your models alike? How were they different?*

> ***Choice 1:*** Represent the number 425 in three different ways. Tell what each way helps you best see about 425.
>
> ***Choice 2:*** Represent the number 310 in a way that makes it easy to see that 310 is greater than 300. Then represent 310 so it is easy to see that it is an even number. Tell why each representation shows these aspects of the number.

CCS: Number & Operations in Base Ten: 2.NBT   **BIG IDEA:** 1.3
Mathematical Practice: 5

It is important that all students realize that different representations of numbers show different things about them. But in ***Choice 2***, the students are guided slightly more heavily than in ***Choice 1***. Some students appreciate a bit more guidance, and this is always a way to differentiate.

A student might show that 425 is less than 500 by placing it to the left of 500 on a number line. He or she might show that 425 is a multiple of 25 by showing $4.25 in quarters. He or she might show that 425 is odd by writing it as 212 + 213 (almost a **double**, but one off).

Questions that might be asked of all students include:

- *How do your representations show that your number is less than 1000? More than 100?*
- *Which representation did you think of first? What did that one show?*
- *Suppose you had used only base ten rods and units to represent your number. What might you have shown about your number?*

Parallel Tasks for Pre-K–Grade 2

TEACHING TIP. One way to create parallel tasks is to ensure that one choice is more guided and one choice more open. In this way, different types of students will be able to find the sort of question they prefer.

Choice 1: The sum of the digits of a three-digit number is 12. How many base ten blocks could you use to model that number?

Choice 2: You can model a number with 58 base ten blocks. What other numbers of base ten blocks could you use to model that number?

CCSS: Number & Operations in Base Ten: 1.NBT, 2.NBT   **BIG IDEA:** 1.4
Mathematical Practice: 5

One of the important principles involved in modeling a number with base ten blocks is that for numbers greater than 10, there is always more than one way to represent the number with blocks. For example, 32 can be represented by 5 blocks (3 tens + 2 ones), 14 blocks (2 tens + 12 ones), 23 blocks (1 ten + 22 ones), or 32 blocks (32 ones). Another important idea is that the minimum number of blocks required to model a number is the sum of the digits of the number (e.g., 3 + 2 = 5 blocks for 32). Both of the task choices above get at these ideas, but at different levels of sophistication.

The alternative numbers of blocks that can be used to represent a number differ by multiples of 9. This is because when trading blocks, 1 large block is lost for each 10 small ones with which it is replaced; the end result is an increase of 9 blocks. Thus, a number that can be represented with 58 base ten blocks might also be represented, for example, with 22 blocks (because 58 − 22 is a multiple of 9).

Common questions could be asked of both groups of students, for example:

- *What number did you use?*
- *How did you know that the number was correct?*
- *Could you have used a different number? How do you know?*
- *How many other blocks could you have used to represent your number?*

➤ **Variations.** For either choice, numbers can be changed to make the tasks more or less challenging. For Choice 1, students can be given the option of working with two-digit or three-digit numbers, and they can be asked to provide more than one solution. For Choice 2, students can be asked to state a generalization that describes their findings.

Parallel Tasks for Pre-K–Grade 2

> **Choice 1:** A number is about 10, but it's not 10. What is the most it might be? What is the least it might be?
>
> **Choice 2:** A number is about 125, but it's not 125. What is the most it might be? What is the least it might be?

CCSS: Counting & Cardinality: K.CC   **BIG IDEA:** 1.4
　　　　Number & Operations in Base Ten: 1.NBT, 2.NBT
　　　　Mathematical Practices: 3, 6

Again, the difference between the two versions is very slight, but it allows students who are working at different mathematical levels to select the choice that is more appropriate for them.

It is important for students to recognize that the word *about* in the question is vague yet can still be meaningful. They will probably find it interesting that the larger the number, the wider the range of values that will be regarded as appropriate estimates. For example, for numbers about 10, the range might be about 8 to 12; for 125, however, the range might be about 115 to 135.

Common questions could be asked of both groups of students, such as:

- *Was your "most" or "least" more than 5 away from your number? Why or why not?*
- *Did you go up the same amount as you went down? Did you have to?*
- *Will everyone in the class have the same answers? Why or why not?*

> **Choice 1:** Use the digits 1, 2, 3, 4, 5, and 6 to make this true:
> □□□ is almost, but not quite, 300 more than □□□.
> Make it true in different ways.
>
> **Choice 2:** Use the digits 1, 2, 3, 4, 5, and 6 to make this true:
> □□□ is more than 450, but □□□ is less than 450.
> Make it true in different ways.

CCSS: Number & Operations in Base Ten: 2.NBT   **BIG IDEA:** 1.5
　　　　Mathematical Practices: 2, 5, 6

Students use slightly different processes to solve the two given choices, but in both cases they compare three-digit numbers. In the first instance, students are likely to make the hundreds digits 3 apart, but then need to figure out that the tens digit in the first number should be just a little less than the tens digit in the second number. In the second case, students will think more about benchmarks, realizing that the hundreds digit of the first number might be 5 or 6, but could be 4.

Appropriate questions for all students include:

- *Could any of your hundreds digits be 1? Which number? When?*
- *Could any of your tens digits be 1? Which number? When?*
- *Would you choose the hundreds digits first or the ones digits? Why?*

Parallel Tasks for Pre-K–Grade 2

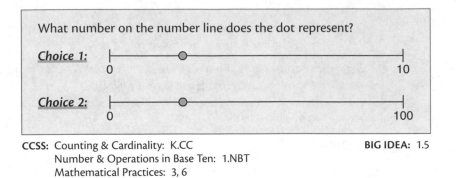

CCSS: Counting & Cardinality: K.CC   **BIG IDEA:** 1.5
Number & Operations in Base Ten: 1.NBT
Mathematical Practices: 3, 6

These parallel tasks allow students who are working with only small numbers to succeed at a task similar to the one used by students able to work with greater numbers. The same strategies are likely to be used no matter which number line is chosen. Most students will mentally divide the line in half, then that half in half, and so on.

Questions that might be asked of both groups of students include:

- *Could your number have been around 25? Explain.*
- *Could your number have ended in a 0? Explain.*
- *Could your number have been close to the number at the end of the line? Explain.*
- *How did you figure out your number?*

> **Variations.** The task can be varied by using alternative endpoints for the number line and alternative positions for the dot.

> _Choice 1:_ To subtract a number from 20, Lia added 4 and then 10. What number did Lia subtract from 20?
>
> _Choice 2:_ To subtract a number from 71, Sara added 3 and then 30 and then 1. What number was she subtracting?

CCSS: Number & Operations in Base Ten: 1.NBT, 2.NBT   **BIG IDEA:** 1.6
Mathematical Practices: 1, 2, 3, 5

An important concept for students to learn is that any subtraction can be solved by using addition, that is, finding the missing amount. It doesn't matter whether the numbers are small or large. In this particular case, students have to put together the pieces to realize what is added, and then realize what subtraction that would relate to.

A number line or a 100 chart might be a useful model. For **_Choice 1_**, students might jump back 10 from 20 and then 4 more to see that 6 was being subtracted. For **_Choice 2_**, students might jump back 1 from 71 and then 30 and then 3 to realize that 37 was the number being subtracted.

Common questions that suit both tasks include:

- *Did you need to figure out the total of what you added or not?*
- *What mathematical tool might be helpful for you to visualize the situation?*
- *Why does it make sense that you could have used addition to figure out the answer to a subtraction problem?*

Choice 1:   You have bunches of bags that hold either exactly 3 jelly beans or exactly 4 jelly beans. You cannot open the bags; you have to use whole bags. What are some numbers you could represent? What are some you could not?

Choice 2:   You have bunches of bags that hold either exactly 6 jelly beans or exactly 9 jelly beans. You cannot open the bags; you have to use whole bags. What are some numbers you could represent? What are some you could not?

CCSS: Number & Operations in Base Ten: 1.NBT, 2.NBT   **BIG IDEA:** 1.7
Mathematical Practices: 1, 2, 4, 7

Decomposition of numbers comes up in many mathematical situations. In this particular problem, students are asked to decompose numbers into either groups of 3s and 4s or groups of 6s and 9s. It provides a lot of opportunity to practice addition. In ***Choice 1***, students might realize that numbers like 3, 4, 7, 10, 11, 13, and so forth are possible, but 1, 2, and 5 are the only impossible numbers. In ***Choice 2***, students might realize that 6, 9, 12, 15, 18, and so forth (i.e., only numbers that are multiples of 3) are possible, but 3 itself is not.

Common questions that suit both tasks include:

- *What are some small numbers you could not represent?*
- *Could you represent 30? How?*
- *Could you represent 20? How?*
- *Are there more numbers you can represent or more that you cannot?*

PARALLEL TASKS FOR GRADES 3–5

What real-life situations might this number describe?

Choice 1:   10,000   **Choice 2:**   1,000

CCSS: Number & Operations in Base Ten: 3.NBT, 4.NBT   **BIG IDEA:** 1.2
Mathematical Practices: 3, 4

Some students in grades 3–5 will find the number 10,000 difficult to relate to. Even if they can read it, they often have no sense of its size. Other students who think more proportionally can imagine 10,000 in terms of known numbers, for example, 10 thousand cubes or the number of people a local arena holds. By providing choices for which number to work with, all students are more likely to benefit from the task.

> *Variations.* A different pair of numbers, whether two other whole numbers, two decimals, or two fractions, can be used instead of the pair above.

TEACHING TIP. One of the benefits of parallel tasks is that even though a student may select an "easier" choice, he or she still benefits from class discussion of the other option.

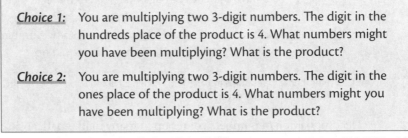

Choice 1: You are multiplying two 3-digit numbers. The digit in the hundreds place of the product is 4. What numbers might you have been multiplying? What is the product?

Choice 2: You are multiplying two 3-digit numbers. The digit in the ones place of the product is 4. What numbers might you have been multiplying? What is the product?

CCSS: Number & Operations in Base Ten: 5.NBT   **BIG IDEA:** 1.3
Mathematical Practices: 6, 7

Clearly, the choices are similar, although *Choice 1* is certainly more challenging than *Choice 2*. But that is what we want, since we want success for all students. In the case of *Choice 2*, a student needs to realize that it is the fact that the product of the units digit ends in 4 that matters and nothing else, so students could, for example, multiply 304 × 101. In the case of *Choice 1*, a student needs to realize that the hundreds place comes from adding digits from a combination of products and perhaps from regrouping as well. A student might choose random values like 312 × 517, notice that the hundreds digit is 3 and then think about whether or how it is possible to increase it to be 4 (312 × 527 works). In either choice, students are thinking about where digits in a multiplication result come from and getting multiplication practice.

Choice 1: Choose digits for the blanks in this subtraction:
☐☐1 – ☐97. Then calculate the difference.

Choice 2: Choose digits for the blanks in this subtraction:
☐☐7 – ☐61. Then calculate the difference.

CCSS: Number & Operations in Base Ten: 3.NBT   **BIG IDEA:** 1.3
Mathematical Practice: 6

Choice 1 would require regrouping if a student uses the **standard algorithm**, whereas *Choice 2* might or might not, depending on what is placed in the second box. However, students using either choice could use various **properties** of operations to make their computation simple. For example, for _**Choice 1**_, a student might choose 421 – 197 and add 3 to 197 to get to 200 and then 221 more, for a total of 224. For _**Choice 2**_, a student might use a similar strategy or might use the standard algorithm (e.g., for 997 – 261) or might use an alternate strategy, such as subtracting in pieces, first the 200, then the 60, and then the 1.

Questions suitable for students using either task include:

- _Which digit did you choose to be greater—the first or the third? Why?_
- _Did you make your second digit a high one or a low one? Why?_
- _Did the digits you chose affect the strategy you used to subtract?_

**Choice 1:**   You round a number to the nearest ten and to the nearest hundred. Both times, the rounded number is less than the original. List three possibilities for the original number.

**Choice 2:**   You round a number to the nearest ten and to the nearest hundred. One rounding increases the number and the other decreases the number. List three possibilities for the original number.

CCSS: Number & Operations in Base Ten: 3.NBT   **BIG IDEA:** 1.4
Mathematical Practices: 3, 6, 7

Some students might realize that a number just a little above a hundred value (e.g., 401) would work for _**Choice 1**_. But a student might choose a value just a little above any multiple of 10, as long as that multiple is less than halfway to the next hundred. For _**Choice 2**_, a student needs to think harder about what kind of number would work. For example, 439 works because the rounded hundred number is less than 439 and the rounded ten number is more than 439. But so does 372, where you go up for the rounded hundred and down for the rounded ten.

Suitable questions for students working on either choice include:

- _If you round to the nearest hundred, what do you know for sure about the rounded value?_
- _What if you round to the nearest ten?_
- _When would a rounded number go up? Go down?_
- _When would a number increase whether you round to the nearest ten or the nearest hundred?_

> **Choice 1:** You add two numbers, and the answer requires more base ten flats than rods and more rods than ones. What numbers could you have added? Explain.
>
> **Choice 2:** You add two numbers, and the answer requires no base ten flats and more rods than ones. What numbers could you have added? Explain.

CCSS: Number & Operations in Base Ten: 3.NBT   **BIG IDEAS:** 1.4, 1.7
Mathematical Practices: 1, 2, 3, 5

By providing information about the answer, but not exact values, these parallel tasks have an open aspect to them. Possible totals for **Choice 1** include 321, 932, 842, etc. Possible totals for **Choice 2** include 72, 91, 50, etc. Students more comfortable working with totals less than 100 might select **Choice 2**.

Once students decide what a possible total is, they still need to break up the number into two addends and should be encouraged to come up with a variety of possibilities.

Some possible questions for both groups of students include:

- *Did you choose the total first or did you choose one of the values first?*
- *Could you have done it the other way?*
- *Could your total have been 135? Why or why not?*
- *Could one of the numbers you added have been 49, even though 49 has more ones than rods?*

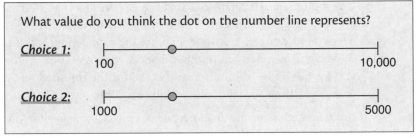

What value do you think the dot on the number line represents?

Choice 1:   100 ——————●———————— 10,000

Choice 2:   1000 ——————●———————— 5000

CCSS: Number & Operations in Base Ten: 4.NBT   **BIG IDEA:** 1.5
Mathematical Practices: 3, 5

These parallel tasks allow some students to work with a smaller number range and smaller numbers if they wish. Students who choose **Choice 2** are likely to mentally divide the line into fourths and see the dot as being at about 2,000. Those who pursue **Choice 1** are likely to estimate the 100 as 0 to help them mentally divide the line.

In discussing the work of all students, a teacher might ask:

- *Is your dot worth more than 1000? Why or why not?*
- *Is it worth more than 2000? Why or why not?*

- *How did you figure out the value the dot represents?*
- *How do you know that there has to be more than one reasonable answer to the question?*

> **Variations.** The task can be varied by using alternative endpoints for the number line and alternative positions for the dot.

Choice 1: You are dividing 4260 by 20. How could you use multiplication to help you divide?

Choice 2: You are dividing 424 by 2. How could you use multiplication to help you divide?

CCSS: Number & Operations in Base Ten: 5.NBT
Mathematical Practices: 3, 7

BIG IDEA: 1.6

No matter how small or large the numbers, division can always be thought of in terms of multiplication. It is important that students consider this not only to check whether answers make sense but also to actually use multiplication as a strategy in order to divide.

Questions like these might be asked of all students:

- *How many digits does your answer have? Why does that make sense?*
- *How could multiplying help you figure out whether it makes sense that 4145 ÷ 15 is a little less than 300?*
- *How did you use multiplication to figure out your division?*

Choice 1: You divide a decimal by a whole number. The quotient is of the form ☐☐.☐☐. The sum of the four digits of the quotient is 8. What division might you have completed?

Choice 2: You divide a four-digit whole number by a two-digit whole number. There is no remainder. The sum of the digits of the quotient is 8. What division might you have completed?

CCSS: Number & Operations in Base Ten: 5.NBT
Mathematical Practice: 6

BIG IDEA: 1.7

Although work with decimal division is part of the Grade 5 Common Core curriculum, some students will need more time than others to be ready to handle those types of questions. To give them more learning time, a parallel task like this one can be used. The students will still be dividing whole numbers, but they will be hearing the responses of the students who did work with decimals. This will give them more exposure to the concept before they are responsible for demonstrating it. In both instances, the relationship between multiplication and division will be highlighted.

Parallel Tasks for Grades 3–5

A possible solution to **_Choice 1_** is 92.48 ÷ 4, although there are many other solutions. A possible solution to **_Choice 2_** is 5,332 ÷ 86.

Questions a teacher might ask could include:

- *Why couldn't the quotient have the digit 9 in it?*
- *How did you know how many digits the whole number part of your dividend had?*
- *How could using multiplication make it easier to solve your division problem?*

Choice 1: You multiply two numbers and the result is 24,000. What numbers might you have multiplied?

Choice 2: You multiply two numbers and the result is 24.00. What numbers might you have multiplied?

CCSS: Number & Operations in Base Ten: 4.NBT, 5.NBT   **BIG IDEA:** 1.7
Mathematical Practices: 2, 7

The two tasks proposed are quite similar in that both involve multiplication and both are dependent on a good understanding of the base ten system. However, one choice involves decimals (although a student might choose to consider 24.00 as a fairly simple whole number) and the other a fairly large whole number.

In both cases, students might use what they know about multiplication by **powers of 10** and what they know about factors of 24. In the first case, they might use combinations like 600×40 or $8 \times 3,000$. In the second case, they might use combinations like 4.0×6.0 (or just 4×6) or 1.5×16.0.

Questions that suit both tasks include:

- *Could either of the numbers be very small?*
- *What is the most one of the numbers could be?*
- *How is knowing the factors of 24 useful?*
- *What other things about multiplying were useful to know to solve your problem?*

Insert each of the digits 0, 1, 2, 3, 4, 5, 6, 7, 8, and 9 in the right spot. Each digit may be used only once.

Choice 1:

$3 \times 5\square = 15\square$

$\square \times \square\square = \square 0$

$6 \times 1\square = 6\square$

$\square \times 1\square = 105$

Choice 2:

$\square \times \square = 5\square$

$4 \times \square = 3\square$

$\square \times \square = 2\square$

$8 \times \square = 4\square$

CCSS: Number & Operations in Base Ten: 4.NBT   **BIG IDEA:** 1.7
Mathematical Practices: 2, 7

Students can choose to work with single-digit products or products involving two-digit numbers, whichever is more suitable for them. For either task, it would be instructive to ask which digits they figured out first and why those digits were the easiest ones to get. Each task is a challenge, just a different sort of challenge. When offered a choice, many more students will have the opportunity for success. (*Solutions:* ***Choice 1***, top to bottom, left to right: 3, 9, 4, 2, 0, 8, 1, 6, 7, 5; ***Choice 2***: 9, 6, 4, 8, 2, 7, 3, 1, 5, 0.)

TEACHING TIP. When numbered choices are offered, the "simpler" option should sometimes be presented as Choice 1 and other times as Choice 2. The unpredictability will ensure that students consider both possibilities when they choose their tasks.

SUMMING UP

The seven big ideas that underpin work in Counting & Cardinality and Number & Operations in Base Ten were explored in this chapter through more than 60 examples of open questions and parallel tasks, as well as variations of them. The instructional examples provided were designed to support differentiated instruction for students at different developmental levels, in this case targeting two separate grade bands: Prekindergarten–Grade 2 and Grades 3–5.

Counting & Cardinality and Number & Operations in Base Ten are fundamental topics in the elementary grades and cover a range of subtopics. It is vital that students meet success in these topics in order to be successful in many other areas of mathematics.

The examples presented in this chapter only scratch the surface of possible questions and tasks that can be used to differentiate instruction in these particular areas of number. Other questions and tasks can be created

---

MY OWN QUESTIONS AND TASKS

Lesson Goal: Grade Level: _____

Standard(s) Addressed:

Underlying Big Idea(s):

_____

Open Question(s):

_____

Parallel Tasks:

Choice 1:

Choice 2:

_____

Principles to Keep in Mind:

- All open questions must allow for correct responses at a variety of levels.
- Parallel tasks need to be created with variations that allow struggling students to be successful and proficient students to be challenged.
- Questions and tasks should be constructed in such a way that will allow all students to participate together in follow-up discussions.

---

by, for example, using alternate operations or alternate numbers or types of numbers (e.g., three-digit instead of two-digit numbers). A form such as the one shown here can be a convenient template for creating your own open questions and parallel tasks. Appendix B includes a full-size blank form and tips for using it to design your own teaching materials.

Number & Operations—Fractions

DIFFERENTIATED LEARNING activities related to fractions and decimals are derived from applying the Common Core Standards for Mathematical Practice to the content goals that appear in the Number & Operations—Fractions domain for Grades 3–5.

TOPICS

Before differentiating instruction in work with fractions and decimals, it is useful for a teacher to have a sense of how fraction and decimal topics develop over the grades.

Prekindergarten–Grade 2

The only work in fractions within the Prekindergarten–Grade 2 grade band appears in the Geometry domain, where students partition shapes into equal-sized pieces and name those pieces by using fraction language.

Grades 3–5

Within this grade band, students work with fractions, beginning with fractions less than one and then moving on to **improper fractions** and mixed numbers. They learn to think of fractions as numbers on number lines, and as equal parts of a whole. In Grade 5, they learn why a fraction can also be thought of as the quotient when the **numerator** is divided by the **denominator**. They compare fractions in a variety of ways, including using **equivalent fractions**.

Students explore addition, subtraction, and multiplication of fractions in real contexts and learn the algorithms for those operations. They also learn how to divide **unit fractions** (those with a numerator of 1) by whole numbers and whole numbers by unit fractions. They learn how decimals are equivalent to fractions with denominators of 10 and 100 and learn about the equivalence of decimals and how decimals are compared. Operations with decimals are covered in Chapter 1 of this resource, in line with the organization of the Common Core Standards.

Grades 6–8

For Grades 6–8, continued work with fractions in covered under The Number System (Chapter 4).

THE BIG IDEAS FOR NUMBER & OPERATIONS—FRACTIONS

In order to differentiate instruction in fractions, it is important to have a sense of the bigger ideas that students need to learn. A focus on these big ideas, rather than on very tight standards, allows for better differentiation.

It is possible to structure all learning in the topics covered in this chapter around these big ideas, or essential understandings:

2.1. Representing a fraction or decimal in an alternate way might reveal something different about that number and might make it easier to compare with other fractions or decimals.

2.2. Fractions and decimals are useful for describing numbers that fall between whole numbers.

2.3. When a fraction or decimal is used to describe part of a whole, the whole must be known.

2.4. There are various strategies for comparing fractions.

2.5. Operations have the same meaning with fractions as they do with whole numbers.

The tasks set out and the questions asked about them while teaching fractions and decimals should be developed to reinforce the big ideas listed above. The following sections present numerous examples of application of open questions and parallel tasks in development of differentiated instruction in these big ideas across the Grades 3–5 grade band.

OPEN QUESTIONS FOR GRADES 3–5

> **OPEN QUESTIONS** are broad-based questions that invite meaningful responses from students at many developmental levels.

> Choose a fraction that is not $\frac{1}{2}$. Represent that fraction in at least three different ways. For each representation of the fraction, describe something that version makes it easier to see about the fraction than other representations do.

CCSS: Number & Operations—Fractions: 3.NF, 4.NF, 5.NF   **BIG IDEA:** 2.1
Mathematical Practice: 5

It is important that students realize that fractions can represent parts of areas, parts of lengths, parts of masses, or parts of capacities or volumes. Fractions are also numbers that happen to be the quotient of the numerator divided by the denominator.

Allowing students to choose their own fractions makes the task more comfortable, although a teacher might encourage some students to pick "unusual" fractions like $\frac{3}{7}$ or $\frac{4}{9}$ and not just familiar ones like $\frac{2}{3}$ or $\frac{3}{4}$.

Discussion might include attention to many important concepts, for example, that in a part-of-set situation, the pieces need not be equal in area or volume; that in a part-of-area situation, the pieces need not be adjacent; that representing 2 of the 3 equal parts in a rectangle is essentially no different from representing 2 of the 3 equal parts in a square, and so forth.

The focus of the discussion should be on what representations show about the fractions being represented. For example, showing $\frac{4}{5}$ as below tends to make it very clear that $\frac{4}{5}$ is $\frac{1}{5}$ less than a whole.

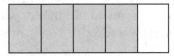

But representing $\frac{4}{5}$ as shown below tends to make it very clear that $\frac{4}{5}$ is 4 groups of $\frac{1}{5}$.

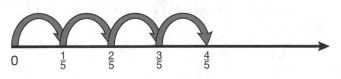

> ➤ **Variations.** Once students are familiar with improper fractions, they might be requested to choose an improper fraction.

> In what situation might you want to use the idea that the fraction $\frac{5}{8}$ could be represented as 0.625? In what situation would you rather call it $\frac{5}{8}$?

CCSS: Number & Operations—Fractions: 5.NF   **BIG IDEA:** 2.1
        Mathematical Practice: 5

Many students believe that it is always easier to work with decimals than with their fraction equivalents. But this is not always the case. For example, if you are trying to determine what $\frac{5}{8}$ of 24 is, you might prefer representing the amount as $\frac{5}{8}$ rather than as 0.625. But if you are comparing $\frac{5}{8}$ to $\frac{3}{5}$, you might prefer comparing 0.625 to 0.6

This question provides an opportunity for students to come up with situations on their own rather than being directed to consider a specific case.

> **Variations.** Fractions other than $\frac{5}{8}$, such as $\frac{2}{3}$, might also be proposed.

> You divide two numbers and the answer is a fraction a little less than $\frac{3}{4}$. What numbers might you have divided?

CCSS: Number & Operations—Fractions: 5.NF   **BIG IDEA:** 2.1
        Mathematical Practices: 5, 6

Most students think of fractions as parts of wholes and that is, of course, useful. But it is even more useful for students to realize that a fraction is a number that could be conceived as the quotient of its numerator divided by its denominator. This question gets to the heart of that notion.

If students really understand this, they might quickly create a fraction slightly less than $\frac{3}{4}$, for example, $\frac{7}{10}$, and say that they had divided 7 by 10.

Alternatively, students might realize that if you divide, for example, 9 by 12 you get $\frac{3}{4}$, so they could divide by something slightly more to get a smaller fraction. In this case they might respond that they divided 9 by 13.

Any two numbers where the first is slightly less than $\frac{3}{4}$ of the second will do.

> **Variations.** The fraction $\frac{3}{4}$ can easily be altered.

> Tell about a time when you would use the number $\frac{1}{2}$.

CCSS: Number & Operations—Fractions: 3.NF   **BIG IDEA:** 2.2
        Mathematical Practice: 4

All students who address this question will think about the fraction $\frac{1}{2}$. Allowing them to come up with their own contexts lets each one enter the mathematical conversation at an appropriate level. Often, when a teacher preselects a context for students, it is meaningful to only some of them.

If students need a stimulus to get started, the teacher can provide drawings that might evoke ideas, for example, a picture showing a sandwich cut in two, a glass half full of juice, or a **hexagonal** table with a line dividing it in half and two people sitting at opposite sides of the table. It is likely that most students will only think of halves of wholes, but some might think of half of a set of two.

TEACHING TIP. Scaffolding tasks by providing models for students reduces the likely breadth of responses that will be offered and may inhibit some students who feel obligated to follow the models.

> A number between 3 and 4 is slightly closer to 4 than to 3. What could the number be?

CCSS: Number & Operations—Fractions: 3.NF, 4.NF, 5.NF   **BIG IDEA:** 2.2
Mathematical Practices: 3, 6

For some students, there are no numbers between 3 and 4. For others, there are a few numbers, like $3\frac{1}{2}$ and maybe $3\frac{1}{4}$ and $3\frac{3}{4}$. This question forces students to think about the **density** of the number line; there are always more numbers, even more fractions, between any two existing ones.

Some students will consider numbers in decimal form, and others in fraction form. Possible responses to the question include 3.51, 3.501, $3\frac{5}{8}$, $3\frac{51}{100}$, and so forth. Students should be required to prove that their values make sense.

➤ *Variations.* Students might be asked to determine a fraction (or decimal) between two fractions (or decimals).

> Draw a small rectangle. Draw a bigger rectangle that the smaller one is part of. Tell what fraction of the big rectangle the small one is.

CCSS: Number & Operations—Fractions: 3.NF   **BIG IDEA:** 2.3
Mathematical Practices: 3, 4

In teaching fractions, students are usually asked to identify a fraction given a partitioned whole. The question above has students thinking the other way—what is the whole if a fractional part is known? Such reversible thinking is an important mathematical process to develop.

To answer a question phrased in this more open way, a struggling student might use a simple fraction such as $\frac{1}{2}$, whereas other students might suggest more complex fractions. Several possibilities are shown in the diagram below:

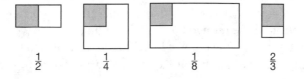

$\frac{1}{2}$ $\qquad$ $\frac{1}{4}$ $\qquad$ $\frac{1}{8}$ $\qquad$ $\frac{2}{3}$

> Describe a situation where $\frac{1}{3}$ is actually more than $\frac{1}{2}$.

CCSS: Number & Operations—Fractions: 3.NF, 4.NF, 5.NF   **BIG IDEA:** 2.3
Mathematical Practices: 3, 6

Normally when we compare fractional values, we assume that the whole is 1. But when a fraction is part of a whole and the wholes are different, a smaller fraction, applied to a greater whole, might in fact be greater. This could be true when talking about portions of objects; for example, $\frac{1}{3}$ of a watermelon might have more volume than $\frac{1}{2}$ of a lemon. But it can also be true for numbers. For example, $\frac{1}{3}$ of 30 is more than $\frac{1}{2}$ of 10.

➤ *Variations.* Instead of phrasing the question as presented above, a teacher might be more direct and ask students to describe a pair of wholes such that $\frac{1}{3}$ of one whole is *more than* $\frac{1}{2}$ of the other, another pair where $\frac{1}{3}$ of one whole is *equal to* $\frac{1}{2}$ of the other, and a third pair where $\frac{1}{3}$ of one whole is *less than* $\frac{1}{2}$ of the other.

> Which two fractions would you find easiest to compare? Why those?
>
> $\frac{4}{5}$ $\frac{4}{10}$ $\frac{2}{9}$ $\frac{1}{12}$ $\frac{8}{9}$

CCSS: Number & Operations—Fractions: 3.NF, 4.NF   **BIG IDEA:** 2.4
Mathematical Practices: 3, 5, 7

This question provides the opportunity to hear what students are thinking about as they compare fractions. Some may only be comfortable comparing fractions with the same denominator; others may be comfortable comparing fractions with the same numerator as well. Yet other students might think about fractions that are close to 0 as compared to those that are clearly closer to 1.

Questions a teacher might ask could include:

- *If you put these fractions on a number line, which would be closest to 0 and which would be closest to 1?*
- *How might you know $\frac{4}{5}$ is more than $\frac{4}{10}$ without even drawing a picture?*
- *Is it possible for $\frac{1}{12}$ of something to actually be more than $\frac{4}{5}$ of something?*

> Choose two fractions with different denominators. Tell how to compare them.

CCSS: Number & Operations—Fractions: 4.NF   **BIG IDEA:** 2.4
Mathematical Practices: 2, 5

Students can use many different strategies to compare fractions. Some students might choose a fraction less than 1 and a fraction greater than 1 by appropriately manipulating the numerators and denominators.

Others might use other benchmarks to make the comparisons simple. For example, they might choose a fraction very close to 0, such as $\frac{1}{1,000}$, and another very close to 1, such as $\frac{99}{100}$. Still others might think of a fraction less than $\frac{1}{2}$, such as $\frac{1}{3}$, and a fraction greater than $\frac{1}{2}$, such as $\frac{3}{4}$.

And other students, perhaps even most students, are likely to select two arbitrary fractions and then use equivalent fractions with a common denominator to compare them.

As students tell how to compare the fractions, they are free to use whatever strategy they wish. This sort of choice is important to allow all students to succeed. It is also helpful to all students, as the comparisons are discussed, to hear about the many different strategies they could have accessed.

> **Variations.** The teacher can require that one or both of the numerators of the two fractions not be 1, because often students tend to use only unit fractions.

> Two fractions are close to 1, but one of the fractions is closer to 1 than the other fraction is. What might the two fractions be and how could you show or tell why the one closer to 1 is indeed closer?

CCSS: Number & Operations—Fractions: 3.NF, 4.NF   **BIG IDEAS:** 2.1, 2.4
        Mathematical Practices: 3, 5, 6

While some students might begin with a picture, others might begin with a symbolic representation. For example, some students realize that $\frac{7}{7} = 1$, so $\frac{6}{7}$ must be close to 1. Similarly, $\frac{5}{5} = 1$, so $\frac{4}{5}$ is close to 1.

The most important part of this question is the explanation: a student describing why, for example, $\frac{7}{8}$ is closer to 1 than, say, $\frac{3}{4}$. Many students are likely to focus on to how far each is from 1 (e.g., $\frac{1}{8}$ away from 1 is closer than $\frac{1}{4}$ away, since $\frac{1}{8}$ is less). But other students will simply appeal to a visual.

> **Variations.** The question could be changed to the fractions needing to be close to $\frac{1}{2}$, or close to 0, for example.

> You multiply two fractions and the result is just slightly less than one of them. What fractions could you have multiplied?

CCSS: Number & Operations—Fractions: 5.NF   **BIG IDEA:** 2.5
        Mathematical Practices: 3, 5, 6

Students with good fraction sense realize that when you multiply by a **proper fraction** less than 1, the result is less than the other factor. To be slightly less, you want to multiply by a proper fraction fairly close to 1. So students should realize that in this case one fraction should be a proper fraction close to 1, such as $\frac{9}{10}$ or $\frac{95}{100}$, and the other fraction can be anything at all.

> You multiply two fractions. The result is $\frac{24}{60}$. What numbers might you have been multiplying? How would you model the multiplication?

CCSS: Number & Operations—Fractions: 5.NF
Mathematical Practice: 5   **BIG IDEA:** 2.5

Although many students will simply look for two numbers to multiply to 24 to be numerators and two to multiply to 60 to be denominators (e.g., creating the fractions $\frac{2}{15}$ and $\frac{12}{4}$), others will simplify $\frac{24}{60}$ to $\frac{2}{5}$ and use $2 \times \frac{1}{5}$, and others will implicitly put conditions on their fractions, for example, deciding both fractions must be proper fractions or in **simplest form**. This allows the question to be appropriately challenging to different groups of students.

Some other possible solutions are $\frac{1}{2} \times \frac{4}{5}$, $\frac{1}{2} \times \frac{24}{30}$, $\frac{2}{3} \times \frac{12}{20}$, etc.

> ➤ *Variations.* Additional problems could be created by changing the resulting value or the operation used.

> You are solving a story problem where the sum is a mixed number a little more than 1. What could the story be?

CCSS: Number & Operations—Fractions: 4.NF, 5.NF
Mathematical Practice: 2   **BIG IDEA:** 2.5

Since the problem requires addition, students must recognize that there has to be some sort of combining. They can then think about what might be combined to make the sum a little more than 1. Many students will first think of two fractions to make exactly 1 and then increase one of them. Possible solutions are $\frac{1}{2} + \frac{5}{8}$, $\frac{9}{10} + \frac{1}{9}$, $\frac{3}{4} + \frac{1}{3}$, etc. They might create stories involving things like pizza, where someone has eaten one fraction of a pizza and then some more, or they might use a more unusual story line.

TEACHING TIP. By asking for values *a little more than, a little less than, close to,* etc., students work a little harder. Usually they first answer the question as though they had to be exact and then they do additional work in figuring out which of their values to change and in which direction.

> You add two fractions and the sum is $\frac{9}{10}$. What could the fractions be?

CCSS: Number & Operations—Fractions: 3.NF, 4.NF
Mathematical Practices: 5, 6   **BIG IDEA:** 2.5

Typically, a teacher provides two fractions and asks students for the sum. If, instead, the sum is given and students are asked for the fractions, a broader range of students can respond successfully.

For example, the objective might be to sum two fractions to get $\frac{9}{10}$. A simple response would be $0 + \frac{9}{10}$. An interesting discussion could ensue as students debate whether 0 does or does not have to be written as $\frac{0}{10}$ to satisfy the instruction or to perform the addition.

Other fairly straightforward solutions are $\frac{8}{10} + \frac{1}{10}$, $\frac{7}{10} + \frac{2}{10}$, and so on. These become more obvious to students if they draw a diagram to represent $\frac{9}{10}$ and simply break up the parts. For example, the diagram below shows how $\frac{9}{10}$ can be represented as $\frac{4}{10} + \frac{5}{10}$:

Looking at the representation, some students might even see $\frac{1}{2} + \frac{4}{10}$ if they recognize that the top row is $\frac{1}{2}$ of the whole and shaded boxes in the bottom row represent $\frac{4}{10}$ of the whole.

Yet other students will work symbolically, either looking for fractions with denominators of 10 to add or by writing $\frac{9}{10}$ as an equivalent fraction, for example, $\frac{18}{20}$ or $\frac{27}{30}$, and adding fractions with denominators of 20 or 30, respectively.

Some students are likely to choose only one pair of addends; others may seek many pairs. In this way, a broader range of students is being accommodated. All of the students will recognize that adding is about putting things together.

> ***Variations.*** It is easy to vary the question by using a different sum or by asking for a particular difference rather than a particular sum.

Create a story problem that you could solve by subtracting $2\frac{1}{3}$ from $4\frac{1}{2}$.

CCSS: Number & Operations—Fractions: 5.NF   **BIG IDEA:** 2.5
          Mathematical Practices: 2, 4

With whole numbers, students learned that subtraction can describe a take-away situation, a comparison situation, or a missing addend situation. The same is true with fractions.

A take-away fraction problem might sound something like this:

I started with $4\frac{1}{2}$ cups of flour but used $2\frac{1}{3}$ cups for a recipe. How much flour do I still have?

A comparison problem involving fractions might sound something like this:

I used $4\frac{1}{2}$ cups of flour and $2\frac{1}{3}$ cups of sugar in my cake. How much more flour than sugar did I need?

A missing addend fraction problem might sound something like this:

I already put in $2\frac{1}{3}$ cups of flour, but I need $4\frac{1}{2}$ cups. How much more do I need to put in?

PARALLEL TASKS FOR GRADES 3–5

PARALLEL TASKS are sets of two or more related tasks that explore the same big idea but are designed to suit the needs of students at different developmental levels. The tasks are similar enough in context that all students can participate fully in a single follow-up discussion.

> _**Choice 1:**_ Two fractions are equivalent. If you add the numerators, the result is 22 less than if you add the denominators. What could the fractions be?
>
> _**Choice 2:**_ Draw a picture to show two equivalent fractions for $\frac{2}{8}$.

CCSS: Number & Operations—Fractions: 3.NF, 4.NF **BIG IDEA:** 2.1
Mathematical Practices: 2, 5

The fact that a part of a whole can be represented in many ways is fundamental to students' ability to add and subtract fractions. Although many students learn the rule for creating equivalent fractions, some use it without understanding why it works; others generalize the rule inappropriately, for example, adding the same amount to both numerator and denominator instead of only multiplying or dividing by the same amount.

The choice of tasks shown for this exercise allows the student who is just beginning to understand equivalence to show what he or she knows about why two fractions might be equivalent. It also allows the more advanced student to work in a more symbolic way to solve a problem involving equivalence. Because there are many solutions to _**Choice 1**_ (e.g., $\frac{2}{4} = \frac{20}{40}$ and $(40 + 4) - (20 + 2) = 22$; $\frac{6}{10} = \frac{27}{45}$ and $(10 + 45) - (6 + 27) = 22$; $\frac{8}{10} = \frac{80}{100}$ and $(10 + 100) - (8 + 80) = 22$), students must recognize that they should try more than one combination of values.

Questions that could be asked of both groups include:

- _What two equivalent fractions did you use?_
- _How did you know they are equivalent?_
- _What kind of picture could you draw to show that they are equivalent?_
- _How did you solve your problem?_

> _**Variations.**_ Instead of talking about the sums of numerators and denominators, Choice 1 could ask for two fractions where the differences between the respective numerators and denominators are 4 in one case and 44 in the other.

Parallel Tasks for Grades 3–5

> ___Choice 1:___ Choose a denominator ☐ that is more than 4. Draw a
> picture that helps explain why $\frac{4}{☐}$ is what $4 \div ☐$ turns out
> to be.
>
> ___Choice 2:___ Choose a denominator ☐ that is less than 4. Draw a
> picture that helps explain why $\frac{4}{☐}$ is what $4 \div ☐$ turns out
> to be.

CCSS: Number & Operations—Fractions: 5.NF   **BIG IDEA:** 2.1
Mathematical Practice: 5

Some students will find it easier to relate an improper fraction to the concept of division. For example, a student might think of $\frac{4}{2}$ as $4 \div 2$ since $4 \div 2$ asks how many 2s are in 4, and that is exactly what you mean when you think of how many wholes 4 halves make.

Other students will be comfortable with proper fractions and thinking of division as sharing. They might think of $\frac{4}{10}$ as the amount an individual would get if 4 items are shared fairly among 10 people. (Each person gets $\frac{1}{10}$ of each of the 4 items, for a total of $\frac{4}{10}$.)

Still other students might be comfortable with proper fractions by thinking of, for example, $\frac{2}{3}$ as asking how much of a 3 fits in a 2 (instead of how many 3s fit in a number).

TEACHING TIP. When working with fractions, switching between proper and improper fractions is often an automatic and valuable way to differentiate.

> ___Choice 1:___ List three fractions between 4 and 5. Explain how you
> know you are right.
>
> ___Choice 2:___ List three fractions between $\frac{4}{8}$ and $\frac{5}{8}$. Explain how you
> know you are right.

CCSS: Number & Operations—Fractions: 3.NF, 4.NF   **BIG IDEA:** 2.2
Mathematical Practice: 5

Some students might be more comfortable determining fractions between 4 and 5 (simply using mixed numbers and then changing them to improper fractions), while others might be just as comfortable using equivalent fractions for $\frac{4}{8}$ and $\frac{5}{8}$ to determine fractions between these two (e.g., $\frac{9}{16}$ or $\frac{45}{80}$).

In either case, it is valuable to reinforce the notion that there are always more fractions that could have been selected.

> **_Choice 1:_** 20 is _____ of _____. Fill in the blanks to make this statement true in four different ways so that the number in the first blank is a fraction with a numerator that is NOT 1 and the number in the second blank is any number you choose.
>
> **_Choice 2:_** 8 is _____ of _____. Fill in the blanks to make this statement true in four different ways so that the numbers in each blank are fractions and the numerators are NOT 1.

CCSS: Number & Operations—Fractions: 3.NF, 4.NF   **BIG IDEA:** 2.3
       Mathematical Practices: 2, 5, 7

Because fractions can be applied to different wholes, students can explore the notion that the same value (whether 20 or 8) can be different fractions of different wholes. For example, 20 is $\frac{2}{10}$ of 100, but it is also $\frac{2}{5}$ of 50. The second choice might be more difficult for some students, but it certainly can be handled by many students. For example, 8 is $\frac{40}{3}$ of $\frac{3}{5}$ or is $\frac{16}{4}$ of $\frac{8}{4}$.

Students who realize that they can apply the routine for multiplying fractions in **_Choice 2_** will probably experience more success.

Not allowing for unit fractions makes the task slightly more challenging.

> **_Variations._** Unit fractions (with numerator 1) could be allowed as fill-ins.

> **_Choice 1:_** Model two fractions with the same numerator. Tell which is greater and why.
>
> **_Choice 2:_** Model two fractions with the same denominator. Tell which is greater and why.

CCSS: Number & Operations—Fractions: 3.NF   **BIG IDEA:** 2.4
       Mathematical Practices: 3, 5

Students are given a great deal of flexibility in deciding which fractions to use. They could use simple denominators such as 2 or 4, or more complex ones. They are free to use proper or improper fractions, whichever they choose. Because many students find it simpler to compare fractions with the same denominator rather than the same numerator, **_Choice 2_** is provided. However, students ready for **_Choice 1_** should be encouraged to take it.

Questions that suit both tasks include:

- *Is either of your fractions really close to 0? Really close to 1?*
- *Is either of your fractions greater than 1? Is the other one?*
- *Do you need to rename the fractions to decide which is greater, or is it easy to tell without renaming?*
- *Which of your two fractions is greater? How could you convince someone?*

TEACHING TIP. Students should occasionally be encouraged to explain why they selected the choices they did.

**Choice 1:** $\frac{\square}{3}$ is more than $\frac{*}{4}$. List four pairs of values for $\square$ and *, but don't use the same number for both $\square$ and *.

**Choice 2:** $\frac{\square}{4}$ is less than $\frac{*}{8}$. List four pairs of values for $\square$ and *.

CCSS: Number & Operations—Fractions: 3.NF, 4.NF   **BIG IDEA:** 2.4
Mathematical Practice: 6

There are many possible solutions for each choice. For example, $\frac{2}{3}$ is more than $\frac{1}{4}$, or $\frac{10}{3}$ is more than $\frac{3}{4}$ for _**Choice 1**_. And $\frac{3}{4}$ is less than $\frac{10}{8}$, or $\frac{9}{4}$ is less than $\frac{30}{8}$ for _**Choice 2**_.

The use of 4ths and 8ths in _**Choice 2**_ might make that task more attractive to some students. They might write $\frac{\square}{4}$ as an equivalent fraction with a denominator of 8 and then adjust the numerator.

Students might use a variety of strategies to compare their fractions and should be encouraged to explain their strategies.

**Choice 1:** You add two fractions. The sum is an improper fraction with a denominator of 20. What might you have added? Think of a few possibilities.

**Choice 2:** You add a fraction with a denominator of 6 to a fraction with a denominator of 9. What could the denominator of the answer be? Why?

CCSS: Number & Operations—Fractions: 5.NF   **BIG IDEA:** 2.5
Mathematical Practices: 1, 3, 5

In both instances, students consider the fact that when we add fractions, we normally use equivalent fractions with a common denominator. In _**Choice 1**_, students are told the new denominator and asked to determine fractions; in _**Choice 2**_, it is the reverse.

Students responding to the first task might consider fractions with denominators that are both 20, or one that is 4 and one that is 5, or one that is 4 or 5 and one that is 20. They also must ensure that the result is greater than 1. Students responding to the second task might realize that, depending on the numerators, the new denominator might be 9 or 18 or even 36.

Questions that suit all students include:

- _Are you free to choose any numerators you want?_
- _What is usually true about the denominators of the fractions you add compared to the denominator of the sum? Does that have to be true?_

> **_Choice 1:_** A rectangle has a length between 3" and 4", but closer to 3". It has a width between 1" and 2", but closer to 2". What could its area be?
>
> **_Choice 2:_** The area of a rectangle is $2\frac{1}{2}$ square inches. What could the length and width be?

CCSS: Number & Operations—Fractions: 5.NF   **BIG IDEA:** 2.5
Mathematical Practices: 4, 6

In either instance, students use the notion that the area of a rectangle is the product of its length and width. In the first situation, the **linear dimensions** are given, but in the second situation, it is the product that is given and students are likely to work backward.

Students can choose the values with which they are comfortable in **_Choice 1_**, for example, $3\frac{1}{4}$ rather than $3\frac{7}{16}$ or $1\frac{3}{4}$ rather than $1\frac{9}{10}$. In **_Choice 2_**, there is still flexibility in choosing two values that multiply to $\frac{5}{2}$ (or an equivalent to $\frac{5}{2}$), for example, $\frac{5}{3} \times \frac{6}{4}$.

Questions that suit all students include:

- *What operations do you use when you are determining the area of a rectangle? Why?*
- *Could the number or numbers you are looking for get really high or not?* [**Note:** A length or width could be high if the other dimension is very small for **_Choice 2_**. The area is limited in **_Choice 1_**.]
- *How could you model that your calculations are correct using grid paper?*

> **_Choice 1:_** You fill 8 drinking glasses $\frac{3}{4}$ full. How many full-to-the-top glasses could have been filled instead?
>
> **_Choice 2:_** You fill some drinking glasses $\frac{5}{6}$ full. If you rearranged the water, you could have filled whole glasses instead with nothing left over. How many glasses could have been filled $\frac{5}{6}$ full?

CCSS: Number & Operations—Fractions: 4.NF   **BIG IDEA:** 2.5
Mathematical Practices: 2, 3, 5

Both of the problems posed involve multiplication of a whole number by a fraction or **repeated addition** of a fraction, but the student must recognize that. In the first situation, the student knows exactly the values he or she has to work with, but the total is not known in **_Choice 2_**, making it more challenging for some students. Working on either problem could help students realize that $c \times \frac{a}{b}$ is a whole number only if b is a factor of $c \times a$.

Whereas there is only one answer to **_Choice 1_**, 6 glasses, there are multiple answers to **_Choice 2_**, including 6 glasses, 12 glasses, 18 glasses, and so forth.

Whichever task is attempted, students could be asked:

- *Could it have been 2 glasses? Why or why not?*
- *Could it have been 3 glasses? Why or why not?*
- *Could it have been 4 glasses? Why or why not?*
- *How could thinking about 3 glasses or 4 glasses have helped solve the problem?*

Choice 1: You divide a fraction of the form $\frac{1}{\square}$ by a whole number. The result is less than 0.1. Describe several possible fraction/whole number combinations and draw pictures to show why you are right.

Choice 2: You divide two whole numbers and the result is a fraction less than 0.1. Describe several possible pairs of whole numbers and draw pictures to show why you are right.

CCSS: Number & Operations—Fractions: 4.NF **BIG IDEA:** 2.5
Mathematical Practice: 5

Some students will gravitate toward **Choice 2** because they are attracted to working with whole numbers. Others will select **Choice 1** because they feel that one of the numbers they need to use is more apparent; it must be of the form $\frac{1}{\square}$.

In the first instance, the students will be looking for a denominator and a whole number with a product greater than 10, so the result will be less than 0.1, or $\frac{1}{10}$ (e.g., $\frac{1}{5} \div 4$). In the second instance, the students will be looking for a numerator that is less than $\frac{1}{10}$ of the fraction's denominator (e.g., $3 \div 50$, or $\frac{3}{50}$). In this case, students must show their understanding that a fraction is the quotient of the numerator and the denominator.

Questions a teacher might ask could include:

- *Could you be dividing by the whole number 2? Explain.*
- *Could you be dividing by a whole number less than 10? Explain.*
- *What kind of picture did you use to show your division? Why that kind?*
- *Once you got a first solution, how did you get a second one?*

Choice 1: You multiply a fraction by $\frac{4}{3}$. What do you know, for sure, about the result?

Choice 2: You multiply a fraction by $\frac{2}{3}$. What do you know, for sure, about the result?

CCSS: Number & Operations—Fractions: 5.NF **BIG IDEA:** 2.5
Mathematical Practices: 2, 3

It is important for students to learn that multiplying by a fraction greater than 1 results in an answer greater than the number being multiplied, but multiplying by a fraction less than 1 results in a smaller answer. There are students who are less comfortable with improper fractions, so ***Choice 2*** is available for them.

In the proposed tasks, no particular numbers to multiply are provided. Students either will have to try a lot of different numbers or will have to think about the general effect of the multiplication.

A particular fraction, with a denominator of 3, was chosen as one of the multipliers to encourage an even greater diversity of responses. For example, some students will assume that if you multiply by thirds, the denominator of the result has to be a multiple of 3 (which may not be true after simplification, but will be true initially) or that if you multiply using a numerator of 2 or 4, the numerator of the result must be even (which also may not be true after simplification).

Questions that might be asked no matter which task is completed include:

- *What sort of model would you use to show the multiplication?*
- *What do you know about the numerator of the resulting fraction?*
- *What do you know about the denominator?*
- *Will your answer be more or less than the number you multiplied by? Why?*

SUMMING UP

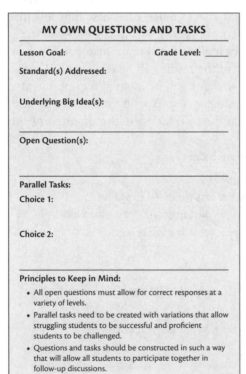

MY OWN QUESTIONS AND TASKS

Lesson Goal: Grade Level: _____

Standard(s) Addressed:

Underlying Big Idea(s):

Open Question(s):

Parallel Tasks:
Choice 1:

Choice 2:

Principles to Keep in Mind:
- All open questions must allow for correct responses at a variety of levels.
- Parallel tasks need to be created with variations that allow struggling students to be successful and proficient students to be challenged.
- Questions and tasks should be constructed in such a way that will allow all students to participate together in follow-up discussions.

The five big ideas that underpin work with fractions and decimals were explored in this chapter through more than 20 examples of open questions and parallel tasks, as well as variations of them. The instructional examples provided were designed to support differentiated instruction for students in the Grades 3–5 grade band.

Fractions are a difficult concept for many students to grasp. It is important that they meet success in understanding fractions to prepare them for later success in the secondary grades.

The examples presented in this chapter are only a few of the possible questions and tasks that can be used to differentiate instruction in work with fractions. Other questions and tasks can be created by, for example, using alternate operations or alternate numbers (e.g., different numerator/denominator relationships). A form such as the one shown here can be a convenient template for creating your own open questions and parallel tasks. Appendix B includes a full-size blank form and tips for using it to design your own teaching materials.

Ratios & Proportional Relationships

DIFFERENTIATED LEARNING activities in **ratios** and proportional relationships are derived from applying the Common Core Standards for Mathematical Practice to the content goals that appear in the Ratios & Proportional Relationships domain for Grades 6–8.

TOPICS

Before differentiating instruction in work with ratios and proportional relationships, it is useful for a teacher to have a sense of how ratio and **proportion** topics develop over the grades.

Prekindergarten–Grade 2

There is very little specific work on ratios and proportional relationships in the primary grades, but there is some informal work in thinking of numbers as groups of another (e.g., 10 as two 5s or 100 as ten 10s), as well as in using measurement **units**.

Grades 3–5

Although there is no specific mention of proportional relationships in this grade band in the Common Core Standards, there is much informal work that involves **proportional thinking** when students work with multiplication and division (where we specifically think of numbers as groups of other numbers) and when students work with fractions (where we indirectly compare numerators and denominators multiplicatively).

Grades 6–8

There is a significant amount of work on ratios and proportional thinking, including rates and percentages, in Grades 6–8.

THE BIG IDEAS FOR
RATIOS & PROPORTIONAL RELATIONSHIPS

In order to differentiate instruction in ratios and proportional relationships, it is important to have a sense of the bigger ideas that students need to learn. A focus on these big ideas, rather than on very tight standards, allows for better differentiation.

It is possible to structure all learning in the topics covered in this chapter around these big ideas, or essential understandings:

3.1. A ratio is a **multiplicative**, not additive, **comparison** between two quantities.

3.2. Many problems are based on multiplicative comparisons and can be solved by representing ratios in alternate ways.

3.3. Proportional relationships can be represented in many ways.

The tasks set out and the questions asked about them, while teaching ratios and proportions, should be developed to reinforce the big ideas listed above. The following sections present numerous examples of application of open questions and parallel tasks in development of differentiated instruction in these big ideas across the Grades 6–8 grade band.

CHAPTER CONTENTS			
Open Questions: Grades 6–8	70	Parallel Tasks: Grades 6–8	77

OPEN QUESTIONS FOR GRADES 6–8

OPEN QUESTIONS are broad-based questions that invite meaningful responses from students at many developmental levels.

> Create a sentence that uses each of the following words and numbers. Other words and numbers can also be used.
>
> *40, percent, most, 80*

CCSS: Ratios & Proportional Relationships: 6.RP   **BIG IDEA:** 3.1
         Mathematical Practices: 1, 2

This open question allows for diverse responses. For example, students might write:

- <u>Most</u> people know that <u>40</u> is 50 <u>percent</u> of <u>80</u>.
- <u>80</u> <u>percent</u> of <u>40</u> is <u>most</u> of the <u>40</u>.
- 90 <u>percent</u> of <u>40</u> and 90 <u>percent</u> of <u>80</u> are <u>most</u> of <u>40</u> and <u>80</u>.

The numbers 40 and 80 were chosen to make the question more accessible. Most students are comfortable with the concept of 50%.

> *Variations.* The words and numbers that are required can easily be changed to allow for other possibilities. Examples of other words are: *ratio, product, added, almost.*

Ratios and rates are often used in sports situations. Describe a sports situation where a ratio might be used. Tell what is being compared.

CCSS: Ratios & Proportional Relationships: 6.RP   **BIG IDEA:** 3.1
      Mathematical Practice: 4

Many students are interested in sports, and so this question should be engaging for them. If there happen to be students not as interested in this context, it makes sense to allow them to talk about ratios in situations of more interest to them, for example, the arts or the environment.

The intention of the question is to help students see how commonplace the use of ratios is. There are many possible solutions, and many students will be very happy to talk about them. These include, for example:

- *Baseball: batting averages*
- *Football: fumbles per touch*
- *Basketball: assist to turnover ratio*

Another pair of numbers have the same relationship as 20 and 50 do. What might those numbers be?

CCSS: Ratios & Proportional Relationships: 6.RP   **BIG IDEA:** 3.1
      Mathematical Practices: 3, 6

When students decide what the relationship between 20 and 50 is, they have many choices, some of which build on the concept of ratio. If, for example, a student chooses two numbers that are 30 apart (e.g., 100 and 130) indeed that person could argue that there is the same relationship, even though it is not multiplicative or built on ratio.

If, however, a student chooses 200 and 500 because in both cases there is 2 of something (10 or 100) compared to 5 of the same thing (10 or 100), a ratio concept has been applied. In fact, students might choose any pair of numbers where the second number is $2\frac{1}{2}$ of the first amount.

But because the question is open, there are many other directions in which a student might go that are worth discussing. For example, a student might choose two numbers where one is less than three times the other, or where one is more than twice the other, or where both are less than 100. Because the question is so open, there should be lots of participation and success.

Open Questions for Grades 6–8

TEACHING TIP. Asking students to reproduce a relationship between two amounts without saying how those amounts are related can always lead to many answers. A teacher might extend this to operations, too (e.g., *Think of pair of numbers to add that is sort of like multiplying* 33 × 66).

Kayleigh has $35. That is a BIG percent of the price of the jacket she wants to buy. What could the jacket cost? But it is a SMALL percent of the price of the tablet she wants to buy. What could the tablet cost?

CCSS: Ratios & Proportional Relationships: 6.RP   **BIG IDEAS:** 3.1, 3.2
          Mathematical Practices: 1, 2, 3

It is important for students to understand that any number, no matter what it is, can be related to any other number by using a percent and that the percent has to change when either number changes.

In this case, a student might decide that 35 is a big percent of 40, specifically, 87.5%, or a big percent of 50, specifically 70%. So the definition of "big" is up to the student. The important thing for a teacher to observe is that the value of the jacket is not more than $70.

If the same value of 35 is a small percent of the cost of the tablet, the tablet should cost more than $70 and likely more than $150. But, again, it is up to the student to make a choice and defend that choice. For example, she or he might say that 35 is a small percent of 200, specifically, 17.5%.

➤ *Variations.* The value of 35 can certainly be changed.

Seth is 8 years old. He is _____% of the height of his older brother.

What might be a reasonable percent? Why? How old do you think his brother would be for that percent?

CCSS: Ratios & Proportional Relationships: 6.RP   **BIG IDEAS:** 3.1, 3.2
          Mathematical Practices: 1, 2, 3

Students might enjoy investigating average heights for various age groups. For example, an 8-year-old boy might be 50" tall. Since a 10-year-old boy might be 56" tall, the percent could be about 89% if Seth's brother is 10. But perhaps Seth's brother is 15 years old and is 67" tall; then the percent would be only 75%.

Fill in values for the blanks to make this statement true:

72 is ____% of ____.

CCSS: Ratios & Proportional Relationships: 6.RP, 7.RP   **BIG IDEA:** 3.2
          Mathematical Practices: 6, 7

Typically, a teacher might ask a student what a given percentage of a given number is, often with a focus on a particular type of percentage, such as a whole number percentage between 0% and 100%, a fractional percentage, or a percentage greater than 100. This more open question allows the student to choose both the percentage and the number of which the percentage is taken, making the calculation as simple as needed.

Many students who would have difficulty with a question such as *What is 120% of 60?* or *What is the number that 72 is 120% of?* might find it much easier to respond with any of the following:

- *72 is 100% of 72.*
- *72 is 72% of 100.*
- *72 is 50% of 144.*

It is likely that in a class of 30 students, many answers will be offered, allowing students the opportunity to consider a wide range of situations.

➤ **Variations.** The number 72 can be changed. Another possibility is to pose a question such as one of these instead: ____ *is* ____ *% of 50* or ____ *is 40% of* ____ .

> Choose a price for four cinnamon buns. Then choose a different number of cinnamon buns and tell how much that new number of buns would cost. Tell how you know you are correct.

CCSS: Ratios & Proportional Relationships: 6.RP, 7.RP   **BIG IDEA:** 3.2
        Mathematical Practices: 1, 2, 3, 4, 5

By allowing students to choose both the price and the number of cinnamon buns to be used, this problem is suitable for a broad range of students.

Some students will select a simple price (e.g., $4), whereas others will select a more complicated price (e.g., $2.79). Some students will select an easy number of buns (such as 8 or 12, multiples of 4), whereas others will select a more complicated number (such as 3, not a multiple of 4).

By allowing—and discussing—many different solutions, each student in the class will be exposed to how to solve many rate problems.

> Choose something you have been wanting to buy that costs more than $50. Imagine you have $30 saved. What discount does the store need to offer before you can afford it?

CCSS: Ratios & Proportional Relationships: 6.RP, 7.RP   **BIG IDEA:** 3.2
        Mathematical Practices: 1, 2, 4, 5

This problem will be motivating to many students because it is personal. Students choose the item to buy, so it can be something of particular interest to them.

By being allowed to select a price that is more than $50 rather than exactly $50, students are provided even more choice. This will allow them to select a value that will make the calculation manageable. For example, a student might choose a price of $60 so that a 50% discount would immediately get to the $30 available.

Another student might select a price of $300. That student would then have to recognize that the $30 saved is only 10% of the price and realize that the discount needs to be 90%. Yet another student might use a value of $120 and apply a 75% discount.

> **Variations.** The values $50 and $30 can be changed to create subsequent problems or to make the problem either simpler or more complex.

Choose fractions for each blank in the sentence below:

J.D. was practicing drawing. He filled _____ of a page in _____ of an hour. How many pages would he fill in 1 hour at that rate?

CCSS: Ratios & Proportional Relationships: 7.RP   **BIG IDEA:** 3.2
          Mathematical Practices: 1, 2, 4, 5

Unit rate is an important concept and is more difficult for students when fractions are involved. When allowed to choose the fractions they use, struggling students can make the problem very simple (e.g., $\frac{1}{2}$ of a page in $\frac{1}{2}$ of an hour) or fairly simple (2 pages in $\frac{1}{2}$ of an hour), but students who are ready for more can make it more complicated (e.g., $\frac{3}{5}$ of a page in $\frac{2}{3}$ of an hour). Students who are capable of a challenge might be encouraged to use these more interesting fractions.

Questions a teacher might ask include:

- *How can you look at your fractions and quickly decide whether the rate will be more or less than 1 page per hour?*
- *Are there different fractions you could have used in the blanks that would result in the same rate for 1 hour? How?*
- *Why is it easier to use a unit fraction (with a numerator of 1) in the second blank than a different fraction?*

> **Variations.** Instead of allowing for any fraction, you might require an improper fraction for the first blank, for example, $3\frac{1}{2}$ pages in $2\frac{1}{3}$ hours.

Create a sentence that includes these words and numbers:

interest, increase, 50, percent, 4

CCSS: Ratios & Proportional Relationships: 7.RP   **BIG IDEA:** 3.2
          Mathematical Practices: 2, 4

To respond to this question, students will likely think about the fact that interest rates are often given in percentages. They might decide that 50% is too high for an interest rate, but then again, they might not. The advantage of using the 50, though, is that 50% is an easy percentage for struggling students to use.

Possible sentences include:

- *If a 4 <u>percent</u> <u>interest</u> rate <u>increased</u> to <u>50</u>%, that would be great if you were getting it, but awful if you were paying it.*
- *If you invest $<u>50</u> at a 4 <u>percent</u> <u>interest</u> rate for a year, you would get $2 at the end of the year and that would <u>increase</u> what you have.*
- *For every $<u>50</u> you invest at 4 <u>percent</u> simple <u>interest</u>, you <u>increase</u> the amount of money you have.*

> You know that 60% of the students in a school are participating in a special fundraiser. If between 200 and 400 students are participating, exactly how many total students might be in the school? How do you know?

CCSS: Ratios & Proportional Relationships: 7.RP　　　　　　**BIG IDEAS:** 3.2, 3.3
　　　　　Mathematical Practices: 1, 2, 4, 5

Some students will realize that since you multiply by 0.6 to determine 60% of a number, you can divide by 0.6 to determine the number that is multiplied. But other students can approach this problem in other ways, for example, by creating a diagram like the one shown, assuming 300 students:

Using the diagram, the student sees that each 10% section is worth 50, and so the whole is worth 500.

This question is made open by allowing for a range of students participating. In that way, each student can choose a value that he or she finds comfortable.

> Think of two **variables** where one doubles when the other doubles. Create a **table of values** and draw a graph that shows that relationship. What do you notice?

CCSS: Ratios & Proportional Relationships: 7.RP　　　　　　**BIG IDEA:** 3.3
　　　　　Mathematical Practices: 3, 4, 7

The heart of proportional thinking is the recognition that when one number is a multiple of another, and you multiply both by the same factor, the relationship between the numbers does not change. For example, since $6 = 2 \times 3$, then 6 is double 3. If you multiply both by 4 (to get 24 and 12), the larger number is still double the smaller. This is actually what we mean when we say two amounts are proportional.

There are many simple examples students might think of to respond to the task. For example, the variables might be number of people and number of eyes. Or they might be number of triangles and number of sides.

In each situation, students will notice that if the table of values goes up by 1s in the left column, the increase in the right column will also be constant. They will also notice that when they graph the relationship, they will see a line going through (0,0).

Number of quarters	Value in cents
1	25
2	50
3	75
4	100

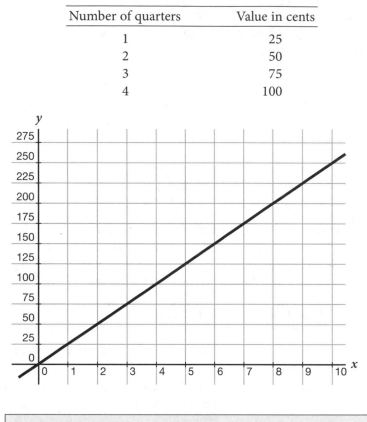

> Jenny is driving VERY FAST. Describe her speed using a variety of units (e.g., miles per hour, yards per minute, inches per second, feet per 5 minutes, kilometers per hour, etc.).

CCSS: Ratios & Proportional Relationships: 7.RP   **BIG IDEA:** 3.3
          Mathematical Practices: 1, 2, 3, 4

Describing a relationship by changing units is a relatively common use of proportional thinking. It emphasizes that the same rate or ratio can be described in a variety of ways. Giving students the choice of units to switch to and giving them a choice about how many different sets of units to use makes this question open and more accessible to more students.

Giving suggestions allows students to see that the equivalent rates need not be unit rates.

PARALLEL TASKS FOR GRADES 6–8

PARALLEL TASKS are sets of two or more related tasks that explore the same big idea but are designed to suit the needs of students at different developmental levels. The tasks are similar enough in context that all students can participate fully in a single follow-up discussion.

> *Choice 1:* The ratio of two numbers is equivalent to 3:4. Could the numbers be 44 apart? Explain.
>
> *Choice 2:* The ratio of two numbers is equivalent to 3:8. Could the numbers be 44 apart? Explain.

CCSS: Ratios & Proportional Relationships: 6.RP   **BIG IDEA:** 3.1
Mathematical Practices: 1, 3, 6

The math in this case focuses on the notion that the multiplicative relationship between two numbers is very different from their **additive relationship**. Even though the difference between the two numbers remains the same, the multiplicative comparisons vary.

In *Choice 1*, the two numbers could be any whole number apart (e.g., 3 and 4 are 1 apart, 6 and 8 are 2 apart, 9 and 12 are 3 apart, . . . , 66 and 88 are 22 apart, . . . , 132 and 176 are 44 apart).

In *Choice 2*, the two numbers must be a multiple of 5 apart (e.g., 3 and 8 are 5 apart, 30 and 80 are 50 apart, 33 and 88 are 55 apart), so they cannot be 44 apart.

Some students might make connections to equivalent fractions.

Questions that suit both tasks include:

- *What are two ratios that are* **equivalent** *to the original ratio?*
- *Could the two numbers that form the ratio be 2 apart? 10 apart?*
- *How did you decide whether 44 was possible?*

> *Choice 1:* The ratio of two whole numbers is 1:2.5. What could the numbers be?
>
> *Choice 2:* The ratio of two fractions is 1:2.5. Both are less than $\frac{1}{2}$. What could the fractions be?

CCSS: Ratios & Proportional Relationships: 6.RP   **BIG IDEA:** 3.1
Mathematical Practices: 1, 3, 6, 7

The student's choice in this set of parallel tasks is whether to work with whole numbers or with fractions. Some students are just more comfortable with whole numbers and will select *Choice 1*. Others might realize that *Choice 2* is easy since you can take any fraction and multiply it by $2\frac{1}{2}$ to get another; you just need to start with a small one to ensure you stay below $\frac{1}{2}$. Possible solutions for **_Choice 1_** are: 2 and 5, 4 and 10, 40 and 100, and so forth. Possible solutions for **_Choice 2_** are: $\frac{1}{10}$ and $\frac{1}{4}$, $\frac{1}{12}$ and $\frac{5}{24}$, $\frac{1}{6}$ and $\frac{5}{12}$, and many others.

In both situations, students might be asked:

- *Is there a smallest number you can use?* [**Note:** The answer is *yes* for **_Choice 1_**, but *no* for **_Choice 2_**.]
- *Is there a largest number you can use?*
- *Is your second number more or less than double your first? Why?*
- *Is your first number more or less than half your second? Why?*

➤ **_Variations._** The ratio can easily be changed to, for example, 2:3.5 or any other whole number and decimal.

Lisa's dad was driving 16 miles every 15 minutes.

Choice 1:   How far would he drive in 20 minutes?

Choice 2:   How far would he drive in 2.5 hours?

CCSS: Ratios & Proportional Relationships: 6.RP, 7.RP   **BIG IDEA:** 3.2
          Mathematical Practices: 1, 2, 4, 5

Writing a rate in another form is the key to solving proportions. In this case, both of the proposed questions involve some thinking. **_Choice 2_** might be perceived as more direct for students who convert the rate to 64 miles per hour and then multiply 2×64 and add 32 miles for 2.5 hours. **_Choice 1_** might be perceived as easier for students who simply add $\frac{1}{3}$ of 16 to 16 to go from 15 minutes to 20 minutes (an additional 5 minutes).

In both situations, students might be asked:

- *Why was it more helpful to know the distance for 15 minutes than for, say, 17 minutes?*
- *What other numbers of minutes would it be very easy to figure out the distance for? Why those amounts of time?*
- *Is it possible to figure out the distance for any number of minutes or only certain numbers?*

TEACHING TIP. It is useful in many problems for students to use "unit rates" that are not 1 minute or 1 hour. Here, for example, a unit of 5 minutes is useful.

> **Choice 1:** Which is a better buy: 5 games for $54.75 or 6 games for
> $63.54?
>
> **Choice 2:** Which is a better buy: 2 games for $21.25 or 6 games for
> $63.54?

CCSS: Ratios & Proportional Relationships: 6.RP, 7.RP   **BIG IDEA:** 3.2
        Mathematical Practices: 1, 2, 4, 5

Both tasks require the use of proportional reasoning. **Choice 1** is suitable for students ready for more of a challenge because the prices are close and there is no simple multiplicative relationship between the numbers 5 and 6. **Choice 2** allows a student to more simply either triple the price for 2 games or take $\frac{1}{3}$ of the price for 6 games in order to compare. Although **Choice 2** is simpler, it still evokes important proportional thinking.

In the first case, the student might calculate the unit cost for each and compare them. Alternately, the student might calculate one unit rate and multiply (rather than divide) by the other number of games to make the comparison, or the student might even multiply the 5-game price by 6 and the 6-game price by 5 to compare the cost of 30 games at both rates.

No matter which task is selected, students might be asked:

- *Is it easy to tell immediately which is a better buy? Explain.*
- *Why would the problem have been easier if both costs had been for the same number of games?*
- *Did you need to know the price for 1 game to solve the problem?*
- *How else could you have solved the problem?*

> **Choice 1:** A number between 20 and 30 is 80% of another number.
>   What could the second number be?
>
> **Choice 2:** A number between 20 and 30 is 150% of another number.
>   What could the second number be?

CCSS: Ratios & Proportional Relationships: 6.RP, 7.RP   **BIG IDEA:** 3.2
        Mathematical Practices: 1, 2, 5

In both situations, students must determine the number of which a given number is part. Rather than choosing a specific number, a range is used. This provides greater practice and allows students to use convenient benchmark numbers if they wish. **Choice 1** uses a percentage less than 100%, whereas **Choice 2** uses a percentage greater than 100%, a more difficult task for most students.

In both instances, students must think about the relationship of the number in the 20s to the other number, that is, whether the other number is greater or less than the given one and why. A possible solution for **Choice 1** is that 24 is 80% of 30; a possible solution for **Choice 2** is that 24 is 150% of 16.

Parallel Tasks for Grades 6–8

Questions that could be asked of both groups include:

- *Is the second number greater or less than the first one?*
- *How did you decide whether the second number was greater or less than the first one?*
- *How far apart are the possible values for the second number?*

> **Variations.** Any percentages under 100% and over 100% might be substituted for 80% and 150%. The whole number range may or may not be changed.

Choice 1: There were 11,544 athletes in the 2016 summer Olympics. Of these, about 5200 were female. Estimate the number and percentage of athletes who were male.

Choice 2: 850 athletes participated in the local track and field event for the Special Olympics. Of these athletes, 512 were female. Calculate the number of athletes who were male. Estimate the fraction who were male.

CCSS: Ratios & Proportional Relationships: 6.RP, 7.RP   **BIG IDEA:** 3.2
Mathematical Practices: 1, 2, 4, 5

Two similar problems are offered, but one involves numbers that may be difficult for some students to work with. By selecting **Choice 2**, students can still show their understanding of when and how to subtract and to use fractions, but in a situation where they are more likely to be successful. When problems are presented in contexts that are of interest to students, it is even more likely that a broad range of students will be engaged.

All students could be asked:

- *How did you calculate the number of males?*
- *Were more than half males? How do you know?*
- *What operation did you use? Why did you use that operation?*
- *How did you estimate? Why was that estimate appropriate?*

Choice 1: How could you use a **double number line** to show how 40% of one number could be equal to 60% of another?

Choice 2: How could you use an **equation** to show how 40% of one number could be equal to 60% of another?

CCSS: Ratios & Proportional Relationships: 7.RP   **BIG IDEA:** 3.3
Mathematical Practices: 4, 5

In **Choice 1**, students use a visual. They might create a number line that shows percentages of the first number and a second "matching" line that shows percentages of the second number.

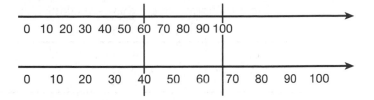

Notice that the visual shows that 67% of the second number matches 100% of the first.

But students who select **_Choice 2_** might write $0.4x = 0.6y$. Therefore, $\frac{y}{x} = \frac{2}{3}$, so the second number is $\frac{2}{3}$ of the first.

Questions appropriate to both groups of students include:

• *How could you have predicted that the first number would be greater than the second?*

• *How does your method make it clear?*

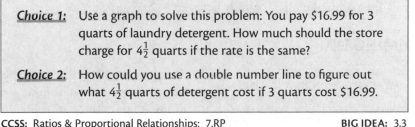

Choice 1:   Use a graph to solve this problem: You pay $16.99 for 3 quarts of laundry detergent. How much should the store charge for $4\frac{1}{2}$ quarts if the rate is the same?

Choice 2:   How could you use a double number line to figure out what $4\frac{1}{2}$ quarts of detergent cost if 3 quarts cost $16.99.

CCSS: Ratios & Proportional Relationships: 7.RP   **BIG IDEA:** 3.3
Mathematical Practices: 2, 5

Both of these situations help students see proportions visually. Students who opt for **_Choice 1_** are likely to draw a graph like this:

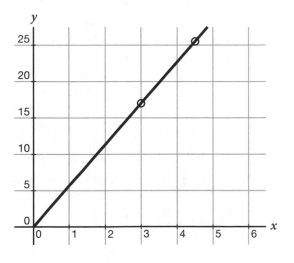

They look for the **_y_-coordinate** on that line where $x = 4.5$.

Students who opt for **_Choice 2_** might draw something like this at the start:

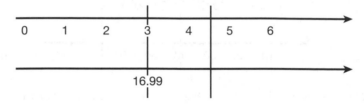

This might be followed by determining the value below 1 (the unit rate) and then repeatedly adding, or perhaps multiplying.

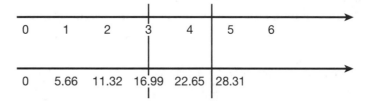

The required price is halfway between $22.65 and $28.31.

SUMMING UP

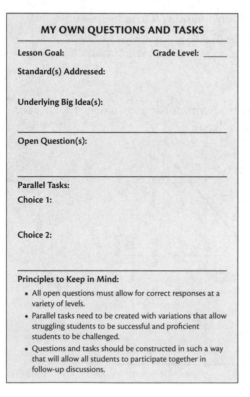

MY OWN QUESTIONS AND TASKS

Lesson Goal: Grade Level: _____

Standard(s) Addressed:

Underlying Big Idea(s):

Open Question(s):

Parallel Tasks:
Choice 1:

Choice 2:

Principles to Keep in Mind:
- All open questions must allow for correct responses at a variety of levels.
- Parallel tasks need to be created with variations that allow struggling students to be successful and proficient students to be challenged.
- Questions and tasks should be constructed in such a way that will allow all students to participate together in follow-up discussions.

The three big ideas that underpin work in ratios and proportional thinking were explored in this chapter through about 20 examples of open questions and parallel tasks, as well as variations of them. The instructional examples provided were designed to support differentiated instruction for students in the Grades 6–8 grade band.

Ratios and proportional thinking are foundational to much of the math we do in our everyday lives, as well as math in the secondary curriculum. Ensuring success in these topics is important.

The examples presented in this chapter are only a sampling of the many possible questions and tasks that can be used to differentiate instruction in ratios and proportional thinking. Other questions and tasks can be created by, for example, using alternate operations or alternate percentages or ratios. A form such as the one shown here can be a convenient template for creating your own open questions and parallel tasks. Appendix B includes a full-size blank form and tips for using it to design your own teaching materials.

The Number System

DIFFERENTIATED LEARNING activities focused on **rational** and **irrational numbers**, including the introduction of **integers**, are derived from applying the Common Core Standards for Mathematical Practice to the content goals that appear in The Number System domain for Grades 6–8.

TOPICS

Before differentiating instruction in work with the number system, it is useful for a teacher to have a sense of how exploration of these topics develops over the grades.

Prekindergarten–Grade 2

Within this grade band, students work only with whole numbers.

Grades 3–5

Fractions and decimals are introduced at this level.

Grades 6–8

In Grades 6–8, there is a significant amount of work with **negative integers** and rational numbers, as well as some work with irrational numbers.

THE BIG IDEAS FOR THE NUMBER SYSTEM

In order to differentiate instruction in the number system, it is important to have a sense of the bigger ideas that students need to learn. A focus on these big ideas, rather than on very tight standards, allows for better differentiation.

It is possible to structure all learning in the topics covered in this chapter around these big ideas, or essential understandings:

4.1. Negative numbers are defined as the **opposites** (based on a line of symmetry at 0) of whole numbers on the number line. This definition leads to the notion that –1 + 1 must be 0.

4.2. Meanings of the operations do not change when dealing with negative integers or rational or irrational numbers, as opposed to whole numbers.

4.3. Properties of the operations do not change when dealing with negative integers or rational or irrational numbers, as opposed to whole numbers.

4.4. Sometimes operating with certain kinds of numbers results in a different kind of number.

4.5. Some numbers cannot be written as either positive or negative fractions, that is, they cannot be written as rational numbers.

While teaching about integers, rational numbers, and irrational numbers, the tasks set out and the questions asked about them should be developed to reinforce the big ideas listed above. The following sections present numerous examples of application of open questions and parallel tasks in development of differentiated instruction in these big ideas across the Grades 6–8 grade band.

OPEN QUESTIONS FOR GRADES 6–8

OPEN QUESTIONS are broad-based questions that invite meaningful responses from students at many developmental levels.

> Two integers are on opposite sides of the number line, but one is just a bit closer to 0 than the other. They are very far apart. What might they be? Which is closer?

CCSS: The Number System: 6.NS   **BIG IDEA:** 4.1
Mathematical Practices: 5, 7

By allowing students to choose how far apart the integers are, instead of telling them, the question becomes more open. Nevertheless, students have to deal with the idea that when integers are on opposite sides of a number line, their distance apart is determined by adding their distances from 0 rather than subtracting them and that integer opposites are equally far from 0.

One student might choose numbers such as –30 and 32, but another might choose –91 and 90. Deciding that –30 is closer to 0 than 32 or that 90 is closer than –91 is an important aspect of the solution.

> *Variations.* Instead of asking for integers that are far apart, the integers could be close together. Instead of using integers, any rational numbers could be used.

> Choose any negative integer. Use two-color counters to represent it in as many ways as you can.

CCSS: The Number System: 6.NS   '   BIG IDEA: 4.1
        Mathematical Practices: 5, 7

A critical piece of understanding for working with integers is the fact that $+1 + (-1) = 0$. Students can represent that fact using as many pairs of opposite color counters as they wish. Allowing them to choose any negative integer gives students some autonomy in the problem. Then, no matter what integer is selected, a student will demonstrate (or not) an understanding that any number of **zero pairs** can be added without changing its value.

> Choose at least one negative value for the blanks to make the following sentence true. Justify your choice.
>
>   ☐ is about 20 less than ☐.

CCSS: The Number System: 6.NS   BIG IDEA: 4.1
        Mathematical Practice: 3

Allowing students to choose their own values can provide a level of comfort. Notice that students could choose two negative values (e.g., –27 and –7), one positive and one negative value (e.g., –15 and 4), or even 0 and a negative value (e.g., –19 and 0). In justifying their choices, students might actually calculate the result, or they might reason without calculating. For example, a student might select –1 and 19, and then explain that she or he started with 1 and 21 (which are 20 apart) and just subtracted 2 from both numbers to get a pair with one negative value.

TEACHING TIP. Use of the word "about" in a question allows for even more freedom in answering correctly.

> You solve a problem that requires you to divide two fractions. The result is slightly less than 1. What might the problem be?

CCSS: The Number System: 6.NS   BIG IDEA: 4.2
        Mathematical Practices: 2, 4

Students need to realize that the two likely situations where they divide fractions involve either determining how many of one amount fit into another or determining a unit rate. For example, $\frac{3}{5} \div \frac{1}{2}$ might ask how many halves are in $\frac{3}{5}$ or might ask for how much of something can be completed in 1 hour if $\frac{3}{5}$ of that thing can be completed in half an hour.

For the given question, students should realize that the answer is slightly less than 1 only if the divisor is slightly more than the dividend.

> **Variations.** The requested solution could be in a different range, for example, slightly more than 2 or slightly less than $\frac{1}{2}$.

> You multiply two integers. The result is about 50 less than one of them. What might the two integers be?

CCSS: The Number System: 7.NS   **BIG IDEAS:** 4.2, 4.3
Mathematical Practices: 1, 5

Students responding to this question must use their multiplication skills as well as their knowledge of sign laws related to multiplying integers. However, there is some latitude in that the difference does not have to be exactly 50, but only about 50. There is also some latitude because students can choose simple factors if they wish.

Possible responses include: −49 and 1, −47 and 1, −2 and 24, but also 0 and 50. Questions a teacher might ask include:

- *Could both of the numbers be positive?*
- *Could both of the numbers be negative?*
- *When is the product of a number less than that number?*

> **Variations.** The question is slightly more challenging if at least one of the integers is required to be negative.

> Two rational numbers that are not integers are added and also subtracted. The sum is 7 less than the difference. What could the numbers be? How do you know?

CCSS: The Number System: 7.NS   **BIG IDEA:** 4.3
Mathematical Practices: 3, 5, 6

Using a number line model helps a great deal with this problem. Students might observe that a sum is only 7 more than a difference if the number that is added or subtracted is $3\frac{1}{2}$.

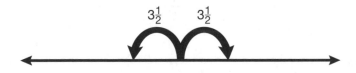

But if the sum is 7 less, as the problem states, it must be $-3\frac{1}{2}$ that is added and subtracted. That is the only way the sum will be less than the difference.

> You add a positive and a negative integer and end up with −20. What could you have added?

CCSS: The Number System: 7.NS
       Mathematical Practice: 3   **BIG IDEA:** 4.4

Students need to realize that to end up at −20, the negative number that was added needs to be −20 or even farther from 0. Any pairs of numbers where the **absolute values** are 20 apart and the number associated with the greater absolute value is negative will work, for example, −21 and +1, or −42 and +22.

➤ *Variations.* The question could be altered so that −20 is replaced by any other positive or negative value.

> Which of these are true? Justify your answer.
> - You can subtract two positive numbers and end up with $-\frac{2}{3}$.
> - You can multiply two negative numbers and end up with $+\frac{4}{5}$.
> - You can divide two positive numbers and end up with $-\frac{2}{3}$.

CCSS: The Number System: 7.NS
       Mathematical Practice: 3   **BIG IDEA:** 4.4

Mathematically, new number systems were created once the result of operations with values within existing systems no longer fit within the system, so this is, historically, a very important big idea. Students have an opportunity in this question to consider "sign laws" involving rational numbers in order to see when the result can or cannot be different from the numbers being operated with.

The question itself is not open, in that there are right and wrong answers, but the justification is open. For example, to show that the first statement is true, students could subtract $\frac{5}{3}$ from 1 or, instead, they could subtract $\frac{8}{3}$ from 2, or $\frac{4}{3}$ from $\frac{2}{3}$.

> An irrational number is close to 4.5. What could it be? List a few possibilities.

CCSS: The Number System: 8.NS
       Mathematical Practices: 3, 5   **BIG IDEA:** 4.5

Students need to know the definition of *irrational number* in order to approach this question. But they have many options for answering it. They might find a **square root** that is close to 4.5 by squaring 4.5 to get 20.25 and then using $\sqrt{20}$. They might choose a **cube root** that is close to 4.5 by cubing 4.5 to get 91.125 and then using the cube root of 91. They might realize that **π (pi)** is an irrational number and use π + 1.4, since π is about 3.1.

The teacher might query students' decisions and ask them to "prove" that their number really is irrational.

> The product of two numbers is irrational and is between 15 and 20. What could the two numbers be?

CCSS: The Number System: 8.NS
Mathematical Practices: 3, 5
BIG IDEA: 4.5

Students need to realize that at least one of the numbers being multiplied (or possibly both of them) must be irrational. Otherwise the product could not be irrational.

Possible pairs of values where both numbers are irrational are, for example, $\sqrt{2}$ and $\sqrt{140}$, or $\sqrt{14}$ and $\sqrt{17}$; possibilities where only one number is irrational include 4 and $\sqrt{16.5}$, or 0.5 and $\sqrt{1000}$.

> You multiply $\sqrt{2}$ by a number and the result is irrational. What could that number have been?

CCSS: The Number System: 8.NS
Mathematical Practices: 7, 8
BIG IDEA: 4.5

Students need to experiment with various multiplications involving rational numbers and irrational numbers to learn that the product of a rational and an irrational must be irrational, whereas a product of two irrationals may or may not be irrational.

This question is very accessible because one possible answer is simply the number 1. However, students should be encouraged to think of a variety of other possibilities. It might be that students choose only whole number values. A teacher could then ask:

- *Could it be a negative number? Which ones?*
- *Could it be a fraction? Which ones?*

> You add two irrational numbers and the answer is around 20. What numbers might you have added?

CCSS: The Number System: 8.NS
Mathematical Practices: 4, 6
BIG IDEA: 4.5

Responding to this question requires students to distinguish between rational and irrational numbers as well as consider the estimated rational values of the irrational numbers. Some students will think that any square root is irrational; this is an opportunity to sort out why $\sqrt{2}$ is irrational, but $\sqrt{4}$ is not. Some students might use the number π or a multiple of π rather than square roots or cube roots. Some students might use an irrational number greater than 20 and add the opposite (the negative) of another irrational number.

Students are also likely to think about **perfect squares**. For example, they might begin by thinking of the numbers 9 + 11, realize that these are $\sqrt{81} + \sqrt{121}$, and then use $\sqrt{80}$ and $\sqrt{120}$.

➤ *Variations.* The number 20 can be varied, as can the stipulation that both numbers are irrational.

PARALLEL TASKS FOR GRADES 6–8

PARALLEL TASKS are sets of two or more related tasks that explore the same big idea but are designed to suit the needs of students at different developmental levels. The tasks are similar enough in context that all students can participate fully in a single follow-up discussion.

> *Choice 1:* A number and its opposite are 24 units apart on a number line. What is the number?
>
> *Choice 2:* A number and its opposite are 5.9 units apart on a number line. What is the number?

CCSS: The Number System: 6.NS
Mathematical Practice: 5   BIG IDEA: 4.1

The two tasks differ only in that one involves the use of integers and the other rational numbers. Both require students to understand the concept of an opposite. One involves a simpler calculation.

No matter which task students select, they could consider:

- *Could one of the numbers be the number in the problem? Why or why not?*
- *Could one of the numbers be double the number in the problem? Why or why not?*
- *What strategy did you use to solve the problem?*
- *How can you check that your answer makes sense?*

> *Choice 1:* List the coordinates for 5 points in the first **quadrant** on a line that does not go through (0,0).
>
> *Choice 2:* List the coordinates for 5 points in at least three quadrants on a line that does not go through (0,0).

CCSS: The Number System: 6.NS
Mathematical Practices: 5, 7   BIG IDEA: 4.1

Some students will be more comfortable staying in Quadrant 1, while others are comfortable using negative coordinates as well. Whichever task students choose, though, they will notice that the increase in the *y*-coordinate divided by the increase in the *x*-coordinate must remain the same for points to be on a line.

Questions suitable for students working on either task include:

- *How do you know that your points are in the quadrant(s) they are supposed to be in?*
- *Did you draw the line first or choose the points first? Could you have done it the other way around?*
- *What do you notice about how far apart the y-coordinates are?*

> **_Choice 1:_**   Show that the product of two numbers can sometimes be greater than the quotient and sometimes less.
>
> **_Choice 2:_**   Choose two numbers to make each statement true:
>
> quotient < difference < sum < product
>
> sum < difference < product < quotient

CCSS: The Number System: 7.NS   Mathematical Practices: 1, 7   **BIG IDEA:** 4.2

Students, and even many adults, harbor many misconceptions about multiplying and dividing. They often believe that multiplication always makes a number greater and division makes a number less, without considering the complication of using fractions or negative numbers. The task shown above requires consideration of those complicating factors.

Choice 1 is less demanding than **_Choice 2_** in that fewer conditions are imposed. As a solution to **_Choice 1_**, a student might offer $-3 \times -4 > -3 \div -4$, but $3 \times 0.5 < 3 \div 0.5$. For **_Choice 2_**, a student might use the pair of numbers 12 and 2 for the first part and the pair of numbers -12 and -0.5 for the second part.

Questions that could be asked of both groups include:

- *Suppose you were using 25 and 5. Which is greater—the quotient or the product?*
- *Is it ever possible for the quotient of two numbers to be greater than the product? When would that be?*
- *What two numbers did you choose? Why did you try those?*

TEACHING TIP. Some students may need to be reminded that they might use proper or improper fractions, as well as decimals and whole numbers, and that positive and negative numbers and 0 might be used.

Parallel Tasks for Grades 6–8

> *Choice 1:* How can you use a model to show what $(-12) \div (-4)$
> means?
>
> *Choice 2:* How can you use a model to show what $(-12) - (-4)$
> means?

CCSS: The Number System: 7.NS   **BIG IDEAS:** 4.2, 4.3
Mathematical Practice: 5

Parallel Tasks for Grades 6–8

Both questions require students to think about what operations with integers mean, in one case division and in the other case subtraction. Some students might appeal to a numerical model, such as a pattern, whereas others might use either a number line model or a counter model.

To interpret $(-12) \div (-4)$, a student might think of this as a measurement (or quotative) division model: *How many groups of (–4) are in a group of (–12)?* Or they might think of this as a missing multiplication, that is, $-4 \times \square = -12$. Notice that the more traditional sharing meaning of division—*How much does each person get if $\square$ people share $\square$ items?*—does not work well for this situation.

To interpret $(-12) - (-4)$, a take-away model works. A student might think about having 12 negative counters and removing 4 of them. On a number line, a student might think of the distance and direction to get from –4 to –12. Or a student might think of a missing addition question: $-4 + \square = -12$.

A discussion of both tasks will elicit valuable conversation that will inform students about even the question they did not answer.

> **Variations.** The problem can be varied by changing the numbers or the operations.

> *Choice 1:* Show –12 as the sum of other integers. What is the
> greatest value you can use for the other integers? What is
> the least value?
>
> *Choice 2:* Show –12 as the product of other integers. What is the
> greatest value you can use for the other integers? What is
> the least value?

CCSS: The Number System: 7.NS   **BIG IDEAS:** 4.2, 4.3
Mathematical Practices: 2, 6

Even though students generally learn about products of integers after learning about sums, many students might find *Choice 2* easier than *Choice 1*. In each instance, students must think about the sign rules that apply to integer operations. They also use reasoning in deciding the greatest and least values that can be used. For example, a student might realize that –12 could be the sum of 400 and –412 or 1000 and –1012, recognizing there is no least or greatest possible value to create a sum. On the other hand, if only integers are used, the least value that can be used to form the product is –12 and the greatest is +12.

Questions that could be asked of both groups include:

- *How did you write −12?*
- *How many combinations did you use?*
- *What was the greatest value you used?*
- *How do you know whether or not a greater value is possible?*

> ***Variations.*** A number other than −12 that has many factors might be used.

Choice 1: Describe three different ways to determine $-\frac{5}{3} \times \frac{3}{8}$.

Choice 2: Describe three different ways to determine $\frac{4}{5} \times \frac{5}{6}$.

CCSS: The Number System: 6.NS   **BIG IDEAS:** 4.2, 4.3
Mathematical Practices: 5, 7

In both cases, students are working with rational numbers, although some students might prefer working with positive proper fractions. Nonetheless, students will show their understanding that calculations involving rationals can be performed in more than one way.

Some students will use decimal or percent equivalents for one or both rational numbers. Some students will recognize that taking thirds of 3 or fifths of 5 is particularly simple and realize that the first expression is equivalent to $-5 \times \frac{1}{8}$ or the second to $4 \times \frac{1}{6}$. Some students will convert $\frac{5}{3}$ to a mixed number and combine one group of $-\frac{3}{8}$ with $-\frac{2}{3} \times \frac{3}{8}$. Some students will draw a model and view that as a different approach.

All students could be asked:

- *Will your answer be positive or negative?*
- *Will it be more than 1 or less than 1?*
- *What do you notice about the numerators and denominators?*
- *Which of your strategies do you find simplest? Why?*

Choice 1: Order these numbers from least to greatest:

$$\sqrt{10}, \ 2\sqrt{3}, \ \sqrt{2} + 2, \ \sqrt{18} - 1$$

Choice 2: Order these numbers from least to greatest:

$$\sqrt{10}, \ \sqrt{15} - 2, \ \sqrt{30} - 3, \ \sqrt{40} \div 3$$

CCSS: The Number System: 8.NS   **BIG IDEA:** 4.3
Mathematical Practice: 6

Both tasks require students to approximate irrational numbers with rational values. The values in **_Choice 1_** are a bit closer to each other than in **_Choice 2_**, since $\sqrt{10}$, $2\sqrt{3}$, $\sqrt{2} + 2$, and $\sqrt{18} - 1$ are approximately 3.16, 3.46, 3.41, and 3.24, but $\sqrt{10}$, $\sqrt{15} - 2$, $\sqrt{30} - 3$, and $\sqrt{40} \div 3$ are approximately 3.16, 1.87, 2.48, and 2.11.

Questions that suit both groups of students include:

- *How did you know that all of the values were greater than 1?*
- *How did you know that all of the values were less than 4?*
- *How might you estimate $\sqrt{10}$ without using a calculator?*
- *Are there two values where you were sure right away that one was probably more than the other? Why?*
- *What was your order?*

➤ **Variations.** As a third option, students might be asked to create a set of four expressions, placed in order, where each involves a different square root and a different operation.

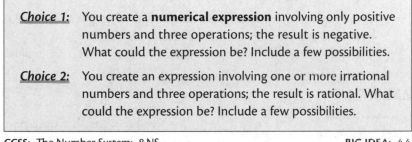

> **Choice 1:** You create a **numerical expression** involving only positive numbers and three operations; the result is negative. What could the expression be? Include a few possibilities.
>
> **Choice 2:** You create an expression involving one or more irrational numbers and three operations; the result is rational. What could the expression be? Include a few possibilities.

CCSS: The Number System: 8.NS   **BIG IDEA:** 4.4
Mathematical Practices: 1, 6, 7

The difference between the two choices is whether students need to work with rational or irrational numbers; many students might still be uncomfortable with irrational numbers and select **_Choice 1_**. Both choices deal with the notion that the result of operations with numbers of one type could be a number of a different type.

In **_Choice 1_**, a student will need to use subtraction. A possibility is $4 \times 3 \div 2 - 80$ or perhaps $1 \times 8 + 6 - 30$. In **_Choice 2_**, a student needs to use irrational numbers that combine, when multiplied or divided, into a rational number. A possibility is $\sqrt{2} \times \sqrt{2} - 0 \div 4$ or $\sqrt{18} \div \sqrt{2} + \sqrt{3} - \sqrt{3}$.

Questions that the teacher might ask all students include:

- *What operations did you include?*
- *Were there any operations you had to include? Why?*
- *How could you have changed your expressions to create even more expressions?*

> **_Choice 1:_** The square root of a fraction is between 3 and 4 and is not a fraction. What could the fraction be?
>
> **_Choice 2:_** The square root of a fraction is between 3 and 4 and is a fraction. What could the fraction be?

CCSS: The Number System: 8.NS   **BIG IDEA:** 4.5
Mathematical Practice: 6

Again, some students might wish to stay in the realm of rational numbers. So a student might use $\sqrt{\frac{49}{4}}$ for **_Choice 2_** or perhaps $\sqrt{\frac{961}{100}}$. But a student selecting **_Choice 1_** might use $\sqrt{\frac{50}{4}}$ or $\sqrt{\frac{950}{100}}$.

Questions that might be asked of all students include:

- *How big did your fraction have to be to have the right size square root?*
- *How did you calculate the square root or estimate it?*
- *How do you know whether the square root is or is not a rational number?*

SUMMING UP

> **MY OWN QUESTIONS AND TASKS**
>
> Lesson Goal: Grade Level: _____
>
> Standard(s) Addressed:
>
> Underlying Big Idea(s):
>
> _____
>
> Open Question(s):
>
> _____
>
> Parallel Tasks:
> Choice 1:
>
> Choice 2:
>
> _____
>
> **Principles to Keep in Mind:**
> - All open questions must allow for correct responses at a variety of levels.
> - Parallel tasks need to be created with variations that allow struggling students to be successful and proficient students to be challenged.
> - Questions and tasks should be constructed in such a way that will allow all students to participate together in follow-up discussions.

The five big ideas that underpin work in the number system were explored in this chapter through about 20 examples of open questions and parallel tasks, as well as variations of them. The instructional examples provided were designed to support differentiated instruction for students in the Grades 6–8 grade band.

The examples presented in this chapter are only a sampling of the many possible questions and tasks that can be used to differentiate instruction in working with integers and rational and irrational numbers. Other questions and tasks can be created by, for example, using alternate numbers. A form such as the one shown here can be a convenient template for creating your own open questions and parallel tasks. Appendix B includes a full-size blank form and tips for using it to design your own teaching materials.

Operations & Algebraic Thinking

DIFFERENTIATED LEARNING activities focused on operations and algebraic thinking are derived from applying the Common Core Standards for Mathematical Practice to the content goals that appear in the Operations & Algebraic Thinking domain for Grades K–5.

TOPICS

Before differentiating instruction in work with operations and algebraic thinking, it is useful for a teacher to have a sense of how the meaning of operations and how algebraic thinking develop over the grades.

Prekindergarten–Grade 2

Within this grade band, students recognize addition as putting together and subtraction as taking apart or taking from. They are introduced to the notion of an equation. They use the relationship between addition and subtraction to solve equations and problems involving whole numbers up to 20. They also begin to work with equal groups to gain initial insights into multiplication.

Grades 3–5

Within this grade band, students represent and solve problems involving multiplication and addition of whole numbers. They explore multiplication and division not only as involving equal groups but also as a way to compare numbers. They use properties of multiplication and division and the relationship between multiplication and division to solve equations and problems. When working with multiplication and division, they particularly explore factors and multiples.

In this band, students also analyze and compare patterns and write and interpret numerical expressions.

Grades 6–8

There are no specific Common Core standards in the domain of Operations & Algebraic Thinking for this grade band, but work involving the number system (see Chapter 4) extends earlier work in operations and algebraic thinking. In addition, work in the Expressions & Equations domain (see Chapter 6) extends work with equations, variables, and so forth.

THE BIG IDEAS FOR OPERATIONS & ALGEBRAIC THINKING

In order to differentiate instruction in operations and algebraic thinking, it is important to have a sense of the bigger ideas that students need to learn. A focus on these big ideas, rather than on very tight standards, allows for better differentiation.

It is possible to structure all learning in the topics covered in this chapter around these big ideas, or essential understandings:

5.1. Decomposing numbers and recomposing them are critical skills for representing, comparing, and operating with numbers.

5.2. Using properties and looking for generalizations that apply to numbers allow us to calculate more fluently and to gain greater insight into numbers.

5.3. It is important to recognize when each operation (addition, subtraction, multiplication, or division) is appropriate to use.

5.4. It is helpful to use the relationships between the operations in computational situations.

5.5. Equations involving unknowns are solved using properties of numbers and meanings of operations.

5.6. Many patterns are built on properties of numbers and operations.

5.7. All patterns are based on **pattern rules**. Without a rule, one cannot be sure how to continue a pattern.

5.8. Patterns can be compared in many ways, including how they grow.

5.9. Patterns can be represented visually to learn more about them.

The tasks set out and the questions asked about them while teaching operations and algebraic thinking should be developed to reinforce the big ideas listed above. The following sections present numerous examples of application of open questions and parallel tasks in development of differentiated instruction in these big ideas across the Prekindergarten–Grade 2 and the Grades 3–5 grade bands.

OPEN QUESTIONS FOR PREKINDERGARTEN–GRADE 2

OPEN QUESTIONS are broad-based questions that invite meaningful responses from students at many developmental levels.

> You add two numbers and end up with more than 7. What could you have added and how do you know? Could one of the numbers have been very small?

CCSS: Operations & Algebraic Thinking: K.OA, 1.OA   **BIG IDEA:** 5.1
        Mathematical Practices: 5, 6, 7

Responses to a question like this one allow the teacher to confirm not only which addition facts students are comfortable using, but also to see whether students realize that many different number combinations work. For example, a student might choose pairs that add to 10, indicating their familiarity with these important combinations. But then they should notice that if one of the numbers was much less than 10, the other was very close to 10, to compensate.

TEACHING TIP. Some students listen to a question and, although they do not follow the instructions to the letter, their answer indicates they have the understanding that was intended to be measured. In the question above, a student might add two numbers and end up with exactly 7. Rather than focusing on the fact that the student did not quite answer the question posed, the teacher should focus on whether the student knew how to add and why one of the numbers could have been very small.

> Choose a number less than 10 and decide how many times you will add it to itself. Make sure the sum is more than 20. What is your number and how many times did you have to add it?

CCSS: Operations & Algebraic Thinking: 1.OA, 2.OA   **BIG IDEAS:** 5.1, 5.2
        Mathematical Practice: 6

Early forays into multiplication often involve repeated addition. In this particular situation, students are free to choose whichever number less than 10 they wish to repeatedly add and also to choose how many times to add it, so long as the answer is above 20.

Examining various solutions will help students see that if a smaller number is chosen, it must be added more times than if a larger number is chosen. This sort of understanding about the effect of unit size will be helpful in measurement situations and helps form the basis of a solid sense of proportional thinking later on.

> *Variations.* Instead of using numbers less than 10 and getting to 20, greater numbers could be used and a greater value exceeded for students ready to focus on multiplication with those types of values.

You subtract two numbers that are close together. What might your answer be?

CCSS: Operations & Algebraic Thinking: K.OA, 1.OA, 2.OA   **BIG IDEAS:** 5.2, 5.3
Mathematical Practice: 3

This question is designed to help students realize that if you subtract two numbers that are close together, no matter how big they are, the difference is small. It helps them think of subtraction more as a difference than only as the result of a take-away.

By encouraging different students to choose different numbers, there will be more data to support the broad idea being examined.

Choose two numbers to add. Use counters to show what the addition would look like.

CCSS: Operations & Algebraic Thinking: K.OA, 1.OA   **BIG IDEA:** 5.3
Mathematical Practice: 5

Asking students to use counters to show what the addition would look like is a way to ensure that the focus is on what addition means and not just on the answer. While some students will probably keep their initial two groups separate, others will simply model the sum. It will then be necessary to go back to ask them what their two numbers were, and it will be interesting to ask whether other pairs of numbers could have been used with this model. The latter question is to focus students on the principle that the grouping can be done in many ways.

> *Variations.* The question could be altered to focus on subtraction instead.

Make up a problem involving both addition and subtraction where the answer is 5.

CCSS: Operations & Algebraic Thinking: K.OA   **BIG IDEA:** 5.3
Mathematical Practices: 2, 4

Students use both their knowledge of the meanings of the operations and their knowledge of facts to respond to this task. There is no restriction on context or on number sizes involved. Students can either state their problem orally, act it out, or draw it.

A sample response might be, *There were 4 kids on the playground. 3 more joined them, but then 2 left. How many kids are still on the playground?*

➤ **Variations.** Alternative versions of this problem might require only one operation and can, of course, require a different result.

> Draw a picture or act out a situation that shows a subtraction problem. Tell what subtraction it shows.

CCSS: Operations & Algebraic Thinking: K.OA   **BIG IDEA:** 5.3
            Mathematical Practices: 2, 4

As important as it might be for students to know subtraction facts, it is equally or even more important for them to know when to apply subtraction. Asking students to create a picture or act out a subtraction situation will demonstrate more clearly whether they know what subtraction means than only approaching the operations symbolically or asking students to look for key words.

Most students will associate subtraction with take away, but it will be interesting to see if any students use other meanings as well, for example, comparison.

Note that no numbers are indicated in the task, allowing students to choose whatever amounts they are comfortable with. Also, while counting is necessary, no actual answer is required, allowing students many comfortable options.

> Describe two different stories that the equation $5 + \square = 9$ could describe and tell how you would solve the equation.

CCSS: Operations & Algebraic Thinking: 1.OA, 2.OA   **BIG IDEAS:** 5.3, 5.5
            Mathematical Practices: 2, 4

Ultimately, mathematics is about de-contextualizing situations to see what the underlying structure is. The equation $5 + \square = 9$ can describe candies or baseball cards or people or money or . . . . But what all of those situations have in common is that there was initially 5 of something and ultimately 9 of that thing. The question is how many were added; that has to be 4, regardless of context.

Asking students to create different stories for the same equation emphasizes the underlying structure we want them to see.

Allowing students to create their own stories also personalizes the problem so that it can be meaningful to them, with contexts that make more sense to them.

➤ **Variations.** Different equations involving different numbers or subtraction instead of addition could bring out the same important big idea.

> You are trying to figure out the unknown number in two different
> subtraction equations. You use a different strategy each time. What
> might the equations have been?

CCSS: Operations & Algebraic Thinking: 1.OA, 2.OA   **BIG IDEA:** 5.4
          Mathematical Practices: 1, 4, 7

Some students might always use the same take-away strategy or perhaps a guess-and-test strategy and might say that they would never use a different strategy. But by asking them to consider this question, they are alerted to at least think about what they actually do.

Other students will use different strategies depending on the numbers used. For example, a student might solve $9 - 1 = ?$ by counting backward from 9, but might solve $9 - 8 = ?$ by counting upward from 8. There could be further discussion about the variety of strategies students do use and why they use different approaches in different situations.

> ➤ **Variations.** The problem could be altered by asking students to consider addition equations instead. Alternately, students could suggest equations that they think they solve in similar ways, even if they involve different operations.

> You know that $4 + 3 = \square$. What other equations have to be true
> if this one is?

CCSS: Operations & Algebraic Thinking: 1.OA   **BIG IDEA:** 5.4
          Mathematical Practices: 2, 3, 7

Some students will simply turn the equation around and say that they know that $\square = 4 + 3$. Other students will use the commutative property and say that this tells them that $3 + 4 = \square$. Still other students will go further and suggest that it tells them that $4 + 4$ must be $\square + 1$ or that $4 + 2$ must be $\square - 1$.

Many students will be quite literal and will simply show a model such as the one below to announce that the unknown number must be 7.

Although such students have not attended to the specific request of discussing other equations, their answers should be accepted as showing an understanding of what the equation means.

TEACHING TIP. It is important for students to be thinking about the fact that, in mathematics, often knowing one thing implicitly tells us many more things too. In some ways, this is the heart of what it means to do math.

Open Questions for Pre-K–Grade 2

> Write three equations involving addition that are true and three equations that are not true.

CCSS: Operations & Algebraic Thinking: K.OA, 1.OA   **BIG IDEA:** 5.5
        Mathematical Practices: 2, 4, 6

Students are free to use any addition combinations with which they are comfortable, but regardless of which combinations are used, the teacher will find out if students understand that an equation represents a balance.

Some students will use simple true equations such as $5 + 1 = 6$, but the teacher should encourage students who are ready to use equations such as $4 + 2 = 5 + 1$ or $4 + 2 = 7 - 1$.

Students should be expected to explain how they know whether the equation is true or not true, ideally by using models of some sort.

> **Variations.** The question can be varied by suggesting that students use subtraction statements instead of addition ones.

> Use the **pan balance**. Put two groups of blocks on each side so that it balances. Write an equation to describe what you did.

CCSS: Operations & Algebraic Thinking: K.OA, 1.OA   **BIG IDEA:** 5.5
        Mathematical Practices: 2, 4, 5

A pan balance is an excellent model for the concept of an equation. It helps students see what it means to say the sides have to be equal.

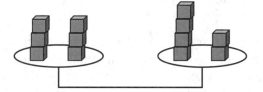

By allowing students to choose how many blocks to put on each side of the balance, a broader range of responses will be achieved, leading to a richer class discussion. Students who wish to use greater numbers are able to, and students still only comfortable with simpler numbers can also be successful.

If possible, the teacher should arrange recording of the various responses in such a way as to encourage generalizations. For example, students might use:

- $6 + 3 = 7 + 2$
- $8 + 3 = 9 + 2$
- $4 + 2 = 5 + 1$

The teacher could bring out the idea that whenever the first number was 1 greater, the second was 1 less and why that is true.

OPEN QUESTIONS FOR GRADES 3–5

Choose a number between 50 and 100. Write that number as equal groups of another number in at least three different ways:

- My number is _____.
- My number is _____ groups of _____.
- It is also _____ groups of _____ and _____ groups of _____.

CCSS: Operations & Algebraic Thinking: 4.OA   **BIG IDEA:** 5.1
Mathematical Practices: 5, 7

Providing counters might be a useful way to help students engage with this open question. The goal is to choose a certain number of counters (between 50 and 100) and arrange them into equal groups. Students are likely to see that certain numbers are easy to express this way (numbers with many factors) while others are more difficult (e.g., 53, a **prime number**).

This question provides an opportunity for students to choose their own values to make sense of ideas about factors and multiples.

Create a line using some orange **Cuisenaire rods**. If you wish, you can add one more rod of a different color to your line. Now create new lines of the same length using rods of only one color. Write equations to describe what you've done.

CCSS: Operations & Algebraic Thinking: 4.OA   **BIG IDEA:** 5.1
Mathematical Practices: 4, 5

Although it may not be obvious at first, students solving this problem are dealing with the notion of factors and multiples. If, for example, a student uses four orange rods to create one line and five brown rods to create another, that individual is actually showing that $4 \times 10 = 5 \times 8$. That is because orange rods are 10 cm long and brown rods are 8 cm long.

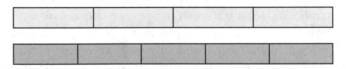

The student, therefore, is learning that 8 is a factor of 40.

By using different combinations of orange rods and rods of other colors, students can explore the factors (up to 10) of any two-digit (or three-digit) number.

> Create a number expression that includes parentheses and at least three operations that would be easier to evaluate if you moved things around before determining the value.

CCSS: Operations & Algebraic Thinking: 5.OA   **BIG IDEA:** 5.1
        Mathematical Practices: 1, 3, 7

One of the Common Core standards for Grade 5 involves evaluating numerical expressions. Ideally, students realize that when evaluating expressions, the expressions should be examined first to see if there are "easier" ways to evaluate them. For example, in earlier grades, students might evaluate $8 + 14 + 2$ by moving the 2 next to the 8 first and then adding the 14.

Possible examples here might include an expression like $2(277 + 328) + 2 \times 23$, where it makes sense to add 277 to 23 and then multiply by 2, and only after that to add 2×328. Another example is $4 \times (5021 + 1843) - 4 \times 21$, where it makes sense to subtract 21 from 5021 and then multiply by 4, and next to add 4×1843.

One of the reasons this question differentiates is because the numbers used can be simple or not simple; there is no expressed requirement.

> You can represent a multiplication using only base ten rods. What numbers might you have multiplied?

CCSS: Operations & Algebraic Thinking: 3.OA   **BIG IDEA:** 5.2
        Mathematical Practice: 5

Students will need to recognize that multiplication involves either repeated addition, equal groups, or an area. The way they arrange their rods will give the teacher insight into what meaning they choose to use.

Realizing that only rods are used, students should recognize that either or both factors must involve a multiple of 10. Some students will realize that some of the rods can be traded for flats and allow for bigger numbers, and others will assume that is not possible and will stick with products like 2×40.

> You multiply a number by another number that is 3 more than your first number. What are some products you can get? What are some products you cannot get? Do you think you can get 100?

CCSS: Operations & Algebraic Thinking: 3.OA   **BIG IDEA:** 5.2
        Mathematical Practices: 3, 6

This question elicits lots of multiplication practice while leaving students with some freedom about what to multiply. The question also helps them to notice patterns. For example, when students compare the answers to $1 \times 4, 2 \times 5, 3 \times 6, 4 \times 7$ to get 4, 10, 18, and 28, they begin to see how the products spread out; this helps build number sense.

Asking students whether they think 100 is possible poses a bit more of a challenge. Some students might simply try combinations, for example, *Does 5 × 8 work? Does 6 × 9 work? Does 8 × 11 work?* Others will notice the pattern of 4, 10, 18, 28, . . . going up by 6, 8, 10, . . . and try to extend the sequence to see if 100 fits the pattern or not. (It does not; there is no 100 possible using whole numbers.)

> You read a numerical expression and SOME of the words you say are:
>
> *multiply by 2, plus, minus, 20*
>
> What might the expression be? Think of a few possibilities.

CCSS: Operations & Algebraic Thinking: 5.OA   **BIG IDEA:** 5.2
        Mathematical Practice: 4

In Grade 5, students are expected to interpret complicated numerical expressions. Rather than simply providing an expression and asking students to interpret it, this more open-ended question encourages students to create their own expression. Possible expressions might be:

$$2 \times (35 + 48) - 20 \quad \text{or} \quad 20 + (39 - 4) \times 2 + 56$$

> About how many days have you been in school? Tell how you estimated. Tell what mathematical operations you used.

CCSS: Operations & Algebraic Thinking: 4.OA   **BIG IDEAS:** 5.2, 5.3, 5.4
        Mathematical Practice: 4

This open question is likely to pique the curiosity of most students. It forces them to consider how they will collect the data and also to consider what computations they will use to answer the question.

Some students will simply focus on the current year. They might think that because there are about 4 weeks in a month and 5 days in most school weeks, they could estimate 20 school days a month. They would then either add or multiply to determine the number of months that have elapsed in the school year.

Others will consider all the years they have been in school. Still others will strive to be more exact, reducing the totals to account for school holidays. Whatever their approach, it is likely that all students will be engaged in the task, and all will be involved in mathematical thinking.

In the discussion of what operations the students used, they should be encouraged to justify why the particular operations were appropriate.

> Draw a picture that shows multiplication by 5.

CCSS: Operations & Algebraic Thinking: 3.OA   **BIG IDEA:** 5.3
   Mathematical Practices: 2, 4

Most students will interpret multiplication by 5 as showing five groups of something or something five times as big as something else. Others might, however, think about groups of 5 and might show, for example, 4 hands, each with 5 fingers, or 6 sets of 5 tally marks.

No matter what picture a student draws, there should be evidence of equal groups either implicitly or explicitly, so that the picture depicts multiplication.

➤ **Variations.** The multiplication need not be multiplication by 5.

> Draw a picture that shows a division.

CCSS: Operations & Algebraic Thinking: 3.OA   **BIG IDEA:** 5.3
   Mathematical Practices: 2, 4

As important as it might be for students to know how to divide, it is equally or even more important for them to know when to apply division. Asking students to create a picture to model a division situation will show more clearly whether they know what division means than only approaching the operations symbolically or having students look for key words.

Most students will associate division with sharing, but it will be interesting to see if any students use other meanings, for example, comparison, as well.

Note that no numbers are indicated in the task, allowing students to choose whatever amounts they are comfortable with. Also, no actual answer is required, allowing students lots of comfortable choices.

➤ **Variations.** Students might be asked to use a specific dividend or divisor instead.

> To describe the total cost in a situation where, for example, you buy an item for \$7 and 3 items for \$4 each, you would write $7 + (3 \times 4)$. Describe a situation where you would write $62 - (4 \times 10)$. Then describe a different situation where you would write $(62 - 4) \times 10$.

CCSS: Operations & Algebraic Thinking: 5.OA   **BIG IDEA:** 5.3
   Mathematical Practices: 2, 4, 6

Students are introduced to the use of parentheses to tell what should happen first, but it is important for them to see *why* parentheses are used. Although this standard will eventually lead to standards that focus on the conventions for the order of operations, at this grade level, students merely get used to the meaning of parentheses. We want students to understand that an expression inside a set of parentheses is a single entity.

> Create a problem you might solve by dividing 60 by a number. Why would that operation help you solve the problem?

CCSS: Operations & Algebraic Thinking: 3.OA   **BIG IDEA:** 5.3
       Mathematical Practices: 2, 4

Division might be applied in situations involving creating equal groups, whether by measuring or not. Students need to know that this is what division is all about, but this question gives them freedom to choose the context they will use and freedom to decide what to divide by. No matter what choices students make, they will still be addressing the issue of when division is applied in real-life situations.

➤ *Variations.* If desired, a specific number can be suggested for the divisor, or a dividend of 24 might be used.

> Create a problem you might solve by using division that results in a remainder of 2. Why does division help you solve the problem? Why is there a remainder?

CCSS: Operations & Algebraic Thinking: 4.OA   **BIG IDEA:** 5.3
       Mathematical Practices: 2, 4

Division might be applied in situations involving creating equal groups, either by measuring or by sharing, but might also be applied in situations involving multiplicative comparisons. For example, one possible response could be, *Kathy had $42. It was $2 more than 4 times as much as Lea had. How much did Lea have?*

Particular meanings of division, such as one based on multiplicative comparisons, could be encouraged, but should not be required, in the interests of differentiation.

➤ *Variations.* Instead of asking about division with a particular remainder, the question could require multiplication where the result is a multiple of 5 or has a tens digit of 4, for example.

> Create a division problem you might solve by subtracting. Tell why the problem is a division problem and why it can be solved by subtracting.

CCSS: Operations & Algebraic Thinking: 3.OA, 4.OA   **BIG IDEA:** 5.4
       Mathematical Practices: 3, 4

Although students usually connect division to multiplication, which is appropriate, they might also connect division to subtraction. No matter which connection is made, students need to know when division applies.

In this particular instance, a student might set up problems like any of these:

- *How many packages of 4 cookies can be made if there are 72 cookies available?* [This can be solved by repeatedly subtracting 4 from 72.]
- *How many pages should Mei read each night if she has to read 72 pages over 4 nights?* [This can be solved by repeatedly subtracting 4 from 72.]
- *Ari has 72 stickers. That's 4 times as many as Alicia. How many stickers does Alicia have?* [This, too, can be solved by repeatedly subtracting 4 from 72.]

Create a pattern that is easy to extend by knowing relationships between multiplying and dividing.

CCSS: Operations & Algebraic Thinking: 4.OA   **BIG IDEA:** 5.6
Mathematical Practices: 3, 7

To respond to this question, a student might realize that multiplying undoes division, so it would be easy to extend any pattern that is built by multiplying by a number to get one term and dividing by the same number to get the next term. An example of such a pattern might be 9, 54, 9, 54, 9, 54, 9, . . . , where the rule is first multiply by 6 and then divide by 6. An alternative might be the pattern 6, 24, 12, 48, 24, 96, 48, . . . , where the rule is first multiply by 4 and then divide by 2. Eventually each new term is double the term two positions before it.

Look at the addition table. List two patterns you see in the table.

+	0	1	2	3	4	5	6	7	8	9
0	0	1	2	3	4	5	6	7	8	9
1	1	2	3	4	5	6	7	8	9	10
2	2	3	4	5	6	7	8	9	10	11
3	3	4	5	6	7	8	9	10	11	12
4	4	5	6	7	8	9	10	11	12	13
5	5	6	7	8	9	10	11	12	13	14
6	6	7	8	9	10	11	12	13	14	15
7	7	8	9	10	11	12	13	14	15	16
8	8	9	10	11	12	13	14	15	16	17
9	9	10	11	12	13	14	15	16	17	18

CCSS: Operations & Algebraic Thinking: 3.OA   **BIG IDEA:** 5.6
Mathematical Practice: 7

With this open question, students are free to decide what patterns to list. They can choose very simple patterns (e.g., noticing that numbers in rows or columns go up by one) or more complex ones (e.g., noticing that the southwest–northeast diagonals are all made up of the same number). Some students might even notice that there are as many 0s as 18s, 1s as 17s, 2s as 16s, and so forth. In asking for only two patterns, the question does not subject struggling students to excessive pressure. In the class conversation, of course, many more than two patterns are likely to come up.

How extensively the teacher discusses the underlying mathematics that explains each pattern depends on what the students are ready for in terms of development and background knowledge.

Look at the multiplication table. List two patterns you see in the table.

×	0	1	2	3	4	5	6	7	8	9
0	0	0	0	0	0	0	0	0	0	0
1	0	1	2	3	4	5	6	7	8	9
2	0	2	4	6	8	10	12	14	16	18
3	0	3	6	9	12	15	18	21	24	27
4	0	4	8	12	16	20	24	28	32	36
5	0	5	10	15	20	25	30	35	40	45
6	0	6	12	18	24	30	36	42	48	54
7	0	7	14	21	28	35	42	49	56	63
8	0	8	16	24	32	40	48	56	64	72
9	0	9	18	27	36	45	54	63	72	81

CCSS: Operations & Algebraic Thinking: 3.OA
Mathematical Practice: 7   BIG IDEA: 5.6

A multiplication table with a 0 row (and 0 column) was used in this example. Many teachers use this table rather than one beginning with 1.

With this open question, students are allowed to decide what patterns to list They can choose very simple patterns (e.g., noticing that numbers in rows or columns go up by the row or column header) or more complex ones (e.g., noticing that every row is either all even numbers or half even numbers).

TEACHING TIP. Many patterns in tables and charts can be explained mathematically. It is valuable to let students try to come up with some of the explanations themselves.

> A pattern begins at 4. Choose an amount to keep adding to get future terms. Tell something you notice about your pattern and why you think what you observed occurs.

CCSS: Operations & Algebraic Thinking: 4.OA   **BIG IDEA:** 5.7
         Mathematical Practice: 7

Students are free to choose any number to add, which makes the question open. What is most interesting, though, is what they notice about the pattern.

For example, a student might choose to add 4 and notice that this produces the *four times table*. Or a student might choose to add 1 and realize that this is the same as just counting, starting at 4. Or a student might choose to add 6 and notice that all of the numbers in the pattern are even. Or a student might choose to add 10 and notice that all of the terms have a ones digit of 4. The important mathematics is in the student's explanation of why these patterns occur.

> **Variations.** Instead of working with an adding pattern, students might start with 400 and keep subtracting to get future terms.

> A certain pattern has the number 5 in it. What could the pattern be?

CCSS: Operations & Algebraic Thinking: 4.OA   **BIG IDEA:** 5.7
         Mathematical Practice: 7

This very open question allows students to make as simple or as complex a pattern as they wish. Many students will simply make a repeating pattern such as 2, 5, 2, 5, 2, 5, . . . ; others might create a pattern with a three-term **core**, for example, 2, 4, 5, 2, 4, 5, 2, 4, 5, ... Still other students will produce an **increasing pattern**, such as 2, 3, 4, 5, ...

No matter what pattern students create, it is important that they be able to respond to a question such as:

- *What makes this a pattern?*
- *Could you change any of the numbers and still have a pattern?*
- *How could you do that?*

> A pattern begins like this: 2, 6, ... How might it continue?

CCSS: Operations & Algebraic Thinking: 4.OA   **BIG IDEA:** 5.7
         Mathematical Practice: 7

There are many possible responses students might come up with. A few are shown below:

       2, 6, 10, 14, 18, . . .   2, 6, 11, 17, 24, . . .
       2, 6, 3, 7, 4, 8, . . .   2, 6, 18, 54, . . .

> **Variations.** This question can easily be adapted by changing the first two numbers shown.

TEACHING TIP. Although there is just as much variability in how to continue patterns where the first three terms are given as in cases where the first two terms are given, it appears that students are much less likely to recognize the variability when the first three terms are given. For example, if they are given 1, 3, 5, most students will continue the pattern by adding 2 to get 7, 9, 11. If they are given 1, 3, students seem more likely to see more alternatives.

> Create two different patterns that begin with the numbers 4, 7, ...
> Tell if, or how, the corresponding terms of the patterns are related.

CCSS: Operations & Algebraic Thinking: 5.OA   **BIG IDEA:** 5.7
        Mathematical Practices: 7, 8

One possible choice might be the patterns 4, 7, 10, 13, 16, . . . , where 3 is added to each term to get the next term, and 4, 7, 11, 16, 22, . . . , where 1 more is added to each new term than was added to the previous term. Corresponding terms are related, although the relationship will be easier for some students to describe than others.

A simple description might be that the terms in the second pattern are always more than the corresponding terms in the first, but increasingly greater as the pattern continues. A more precise comparison might suggest that the corresponding terms in the two patterns differ by 0, 0, 1, 1 + 2, 1 + 2 + 3, and so forth. So, for example, in the second pattern, term 40 is 1 + 2 + 3 + . . . + 38 more than term 40 in the first pattern.

> The terms of one pattern are always 1 more than double the corresponding terms of another pattern. What could the two patterns be? How would their graphs compare?

CCSS: Operations & Algebraic Thinking: 5.OA   **BIG IDEA:** 5.8
        Mathematical Practice: 7

Students pursuing this question have latitude in the sizes of numbers they use. They might use simple patterns such as 1, 2, 3, 4, 5, . . . and 3, 5, 7, 9, 11, . . . or patterns with larger numbers or even decimals, for example, 4.5, 5, 5.5, 6, 6.5, . . . and 10, 11, 12, 13, 14, ...

When comparing the graphs, no matter which values are chosen, the slope of the second line will be greater than the slope of the first one, assuming the first pattern is an arithmetic pattern increasing in a constant fashion. Students might also notice that the increase in the second pattern is always double the increase in the first one.

Other questions that might be asked include:

- *Which pattern could have even numbers? Could both?*
- *Which pattern grows faster? How do you know?*

> List the first 5 numbers in a pattern of your choice. Describe the pattern rule. Then create another pattern using a different rule that is a lot like the first one. Could you predict the 30th number of one pattern by knowing the 30th number of the other?

CCSS: Operations & Algebraic Thinking: 5.OA   **BIG IDEA:** 5.8
Mathematical Practice: 7

Asking students to compare patterns is a valuable way for a teacher to gain insight into student understanding of the properties of patterns. One possibility is to start with a pattern like 4, 6, 8, 10, 12, . . . , where the pattern rule is "Start with 4 and keep adding 2." A similar pattern might be created by changing the start number but not the growth, or by changing the growth and not the start number.

Sometimes it will be easier than at other times to predict the 30th term of one pattern by knowing the 30th term of the other. For example, if the patterns are:

- 4, 8, 12, 16, . . . and 5, 9, 13, 17, . . . , the student could simply add 1 to the term in the first pattern to get the corresponding term in the second pattern
- 2, 4, 6, 8, . . . and 4, 8, 12, 16, . . . , the student could simply double the term in the first pattern to get the corresponding term in the second pattern

Some students might challenge themselves more. For example, with the patterns

- 4, 8, 12, 16, . . . and 6, 12, 18, 24, . . . , students might observe that you could multiply the term in the first pattern by $\frac{3}{2}$ to get the term in the second pattern
- 4, 8, 12, 16, . . . and 4, 7, 10, 13, . . . , some students might observe that you could take $\frac{3}{4}$ of the number in the first pattern and add 1 to get the term in the second pattern

> Draw pictures that might help someone predict the next four terms of the pattern 1, 4, 9, 16, ...

CCSS: Operations & Algebraic Thinking: 4.OA   **BIG IDEA:** 5.9
Mathematical Practices: 5, 7

Many students will draw a representation of the pattern, but some representations may be more useful than others in predicting subsequent terms of the pattern. Two valuable visual approaches to the pattern are shown at the top of the next page.

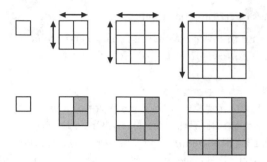

The first set of pictures emphasizes that the numbers are the products of equal values, whereas the second set of pictures emphasizes that the values increase by increasing odd numbers. It is useful for students to see how different representations of a pattern make different aspects of the pattern evident.

> *Variations.* Another pattern that can be used to accomplish the same goal is the pattern 1, 3, 6, 10, ... A third one is a simple growing pattern such as 2, 5, 8, ... In the last case, a student might represent each term as a set of 3s with one item missing in the last group; other students might represent the pattern by emphasizing the start of 2 and a pattern based on adding 3:

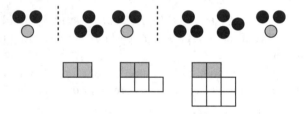

PARALLEL TASKS FOR PREKINDERGARTEN–GRADE 2

PARALLEL TASKS are sets of two or more related tasks that explore the same big idea but are designed to suit the needs of students at different developmental levels. The tasks are similar enough in context that all students can participate fully in a single follow-up discussion.

**Choice 1:**   Two numbers that are not close together add to 9. What could the two numbers be?

**Choice 2:**   Three numbers that are not close together add to 20. What could the three numbers be?

CCSS: Operations & Algebraic Thinking: K.OA, 1.OA   **BIG IDEA:** 5.1
Mathematical Practice: 5

In both instances, students are decomposing numbers, in one case 9 and in one case 20. Rather than asking for all possibilities or asking for particular ones, a condition is used. This requires students to think about what "not close together" means (which is, of course, a matter of opinion). For **_Choice 1_** students are likely to use numbers such as 8 and 1 or maybe 7 and 2. For **_Choice 2_**, students are likely to use numbers such as 14, 1, and 5, or 13, 1, and 6.

Questions suitable to both tasks include:

- *What is the highest any of your numbers could be? Why?*
- *What is the least any of your numbers could be? Why?*
- *How did you choose your first number?*
- *How did you decide whether numbers are far apart or not?*

Choice 1: Take 10 counters. Divide them into 2 piles so that one pile has 4 more counters than the other. How many are in each pile?

Choice 2: Take 10 counters. Divide them into 2 piles so that the piles are close in size. How many are in each pile?

CCSS: Operations & Algebraic Thinking: K.OA
Mathematical Practices: 4, 5

BIG IDEA: 5.1

Both of these tasks require students to decompose the number 10. But rather than telling the students one of the numbers to use, a condition is provided. Notice that the condition in **_Choice 1_** requires more analysis than the one in **_Choice 2_**, and is perhaps better suited to some students than others.

The only possible response for **_Choice 1_** is piles of 7 and 3. It would be interesting to observe students working on this task to see whether they just try all possibilities randomly, or try something and adjust appropriately, moving counters to make the difference closer to 4. Some students might choose to think out or write out all the possible combinations for 10 and look for the one where the two numbers are 4 apart.

Many students will misunderstand the problem and assume that 4 has to be one pile size; they will soon see that this does not work.

There are more options for **_Choice 2_**, although most students will choose 4 and 6 or 5 and 5.

Questions that suit both tasks include:

- *Could either pile have all 10 counters?*
- *Could you have one really big pile and one really small one?*
- *Could the piles be 6 and 5? Why not?*
- *Is there more than one possible answer? Why or why not?*

➤ **_Variations._** The conditions about the pile sizes can easily be adjusted.

> **Choice 1:** Arrange 17 counters so that it is easy to see that 17 is an odd number.
>
> **Choice 2:** Arrange 8 counters so that it is easy to see that 8 is an even number.

CCSS: Operations & Algebraic Thinking: 2.OA   **BIG IDEAS:** 5.1, 5.2
Mathematical Practice: 5

Many students define whether numbers are even or odd based on the digits they end in, but not based on the notion that even numbers can be partnered (which means they are multiples of 2) and odd numbers have a remainder of 1 when partners are created.

Knowing this allows students to respond to this question. Students can show 17 as 8 pairs and 1 more or 8 as 4 pairs.

The difference between the two tasks is both size of number and whether the visual needs to show an even or an odd. Many students would prefer using smaller numbers and are more comfortable showing "evenness" than "oddness."

Questions that could be asked students who have worked on either task include:

- *Could using 10-frames help you do this task? How?*
- *Do you have to count to know whether the number of counters here is even or odd?*

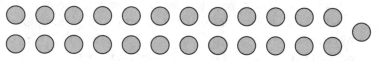

- *How did you show what your problem asked?*

> **Choice 1:** Choose a number to subtract from 20. Draw two different pictures that would help you figure out the result.
>
> **Choice 2:** Choose a number to subtract from 5. Draw two different pictures that would help you figure out the result.

CCSS: Operations & Algebraic Thinking: K.OA, 1.OA   **BIG IDEAS:** 5.1, 5.3, 5.4
Mathematical Practice: 5

Asking students to draw pictures to prove that a certain result is achieved is a way of determining whether they know what subtraction is all about. The two choices here are very similar, but one number is somewhat greater than the other and might be less accessible for some students.

In both cases, students have the opportunity to show whether they think of subtraction as take-away, as comparison, or as a missing addend situation (i.e., what to add to a number to get to 5 or 20).

Questions suitable to both tasks include:

- *How did you decide what number to choose?*
- *Could you have predicted about how much your answer would be?*
- *How do the pictures show the subtraction?*
- *How are your pictures alike? How are they different?*

> **Variations.** You could vary the problem by changing either the numbers involved or the operation involved.

Choice 1: Make up a story problem you could solve by adding 14 and 3. What makes it an addition problem?

Choice 2: Make up a story problem you could solve by subtracting 3 from 14. What makes it a subtraction problem?

CCSS: Operations & Algebraic Thinking: 1.OA, 2.OA   **BIG IDEA:** 5.3
Mathematical Practices: 2, 4

Many students struggle more with subtraction than with addition. By allowing students the choice, those who are tentative with subtraction do not feel the pressure they might if they were required to deal with the subtraction question. However, they still learn about or benefit from the discussion of subtraction that occurs when all students, including those who selected **Choice 2**, talk about their work.

Observing student work will provide insight into what meanings students attach to addition and subtraction.

Questions suitable to both tasks include:

- *What was your story about? Who were the characters?*
- *What did the number 14 represent in your problem?*
- *What did the number 3 represent in your problem?*
- *Could you have predicted whether a solution to the problem would be more or less than 14?*
- *What about your problem allowed you to use the right operation?*

Choice 1: Jane is 8 and her Mom is 37. How old will her Mom be when Jane is 17?

Choice 2: Jane's Mom is 37. How long will it be until she is 50?

CCSS: Operations & Algebraic Thinking: 2.OA   **BIG IDEA:** 5.3
Mathematical Practices: 2, 4

Both choices require students to mathematize a real-life situation. The first problem is slightly more complex than the second simply because there is an extra step.

Questions that could be asked of students working on either task include:

- *Can you just add the numbers in the problem? Why or why not?*
- *What operation or operations did you use to solve the problem? Why those?*
- *Could you have predicted whether your answer would be more or less than 50? Explain.*
- *How did you figure out your answer?*

PARALLEL TASKS FOR GRADES 3–5

> **_Choice 1:_** You multiply two numbers and the answer is in the 20s. What could you have multiplied?
>
> **_Choice 2:_** You multiply two numbers and the answer is 24. What could you have multiplied?

CCSS: Operations & Algebraic Thinking: 3.OA   **BIG IDEA:** 5.1
Mathematical Practices: 5, 6

Both of these tasks encourage students to think of multiplication, but one is more specific than the other and might appeal more to students who like a definitive starting point.

It would make sense to provide students with counters or, if desired, a multiplication table to help them model the multiplications.

Questions that suit both tasks include:

- *Could 0 be one of the numbers? Why or why not?*
- *Could 10 be one of the numbers? Why or why not?*
- *Could the two numbers be the same? Explain.*
- *What do you know for sure about the numbers?*

> **_Choice 1:_** Are there more multiples of 3 or more multiples of 4 between 1 and 100? How many more?
>
> **_Choice 2:_** Are there more prime numbers or more **composite numbers** between 1 and 100? How many more?

CCSS: Operations & Algebraic Thinking: 4.OA   **BIG IDEA:** 5.1
Mathematical Practices: 3, 5

Exploring how numbers can be decomposed into factors is part of the curriculum in Grade 4.

Some students at this level still struggle to decide whether a number is prime. These students might be more successful with **_Choice 1_**, yet they would benefit from the ensuing discussion that involves **_Choice 2_**. For either task, students will make two lists of numbers and compare their lengths. In addition, they will need to justify why particular numbers are on one list, the other list, or both lists (in the case of _**Choice 1**_). As the discussion continues, it would be useful to ask why numbers appear on both lists in _**Choice 1**_ but not in _**Choice 2**_. There are $33 - 25 = 8$ more multiples of 3 between 1 and 100 than multiples of 4. There are $74 - 25 = 49$ more composites than primes. The number 1 is neither prime nor composite.

**Choice 1:** $\square \times \square$ = 480. How many pairs of numbers can you use to fill in the blanks? What are they?

**Choice 2:** $\square \times \square$ = 24. How many pairs of numbers can you use to fill in the blanks? What are they?

CCSS: Operations & Algebraic Thinking: 3.OA, 4.OA   **BIG IDEAS:** 5.1, 5.2
Mathematical Practice: 6

To solve both missing multiplication problems, students could divide the product by potential factors. If students know multiplication factors that result in 12, 24, or 48, this should help them solve the problem. The difference in difficulty focuses primarily on the size of the product. In each case, however, students are encouraged to come up with as many possibilities as they can.

Solutions to _**Choice 2**_ involving whole numbers are 1×24, 2×12, 3×8, and 4×6, but some students might also consider fraction responses such as $\frac{1}{2} \times 48$. Solutions to _**Choice 1**_ involving whole numbers are 1×480, 2×240, 3×160, 4×120, 5×96, 6×80, 8×60, 10×48, 12×40, 15×32, 16×30, and 20×24. But, again, there is an infinite number of other responses if students choose to use fractions.

Questions suiting all students include:

- _Why might it be useful to know that one of the factors must be even?_
- _Why might it be useful to know factors of 12?_
- _Why might it be useful to know factors of 24?_
- _How can you get a second answer once you have a first one?_
- _Are there answers involving fractions? What is one possibility?_

**Choice 1:**   Draw one picture that shows why $4 \times 6 = 6 \times 4$.

**Choice 2:**   Draw one picture that shows why $3 \times 5 = 2 \times 5 + 5$.

CCSS: Operations & Algebraic Thinking: 3.OA   **BIG IDEA:** 5.2
Mathematical Practices: 5, 7

The two proposed tasks encourage students to think about what the properties of multiplication mean. Representing those properties gives a better indication of their understanding of the properties than just citing them.

There are many possible pictures students can draw. For _**Choice 1**_, many students might draw a 4×6 **array**. Even if this is drawn, they still need to explain how it shows why 4×6 is the same as 6×4. But they might draw something else, such as the picture shown below, where there are 6 groups of 4 circles, but also a white, a light gray, a medium gray, and a darker gray group of 6 circles each.

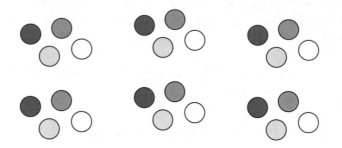

For **_Choice 2_**, students might draw anything that shows that 2 groups of 5 and another group of 5 makes 3 groups of 5. They might, for example, draw 3 hands, with 5 fingers on each.

Questions that suit both tasks include:

- *How many items are in your picture? Why that number?*
- *How did the idea of multiplication show up in your picture?*
- *Why might you want to know that what you modeled is true?*

Which is the lie?

Choice 1:   If you multiply by 5, the ones digit cannot be 4.

If you multiply by 6, the ones digit cannot be 8.

If you multiply by 3, the ones digit could be even or odd.

Choice 2:   If you multiply by 5, the tens digit can be 4.

If you multiply by 2, the ones digit and the tens digit might be the same.

If you multiply by 6, the ones digit can be 5.

CCSS: Operations & Algebraic Thinking: 3.OA   **BIG IDEA:** 5.2
Mathematical Practice: 7

Both of these tasks encourage students to observe patterns and to come to generalizations about multiplication. The format is fun, since most students like to find the lie.

Questions that suit work on both tasks include:

- *What multiplications did you do to make your decision?*
- *Was it easy to eliminate some of the possibilities right away? Which ones?*
- *If a statement says that something can happen and it doesn't happen the first three times you try, can you decide it can't happen?*

> _**Choice 1:**_ Write as many equations as you can to describe this array.
>
>   x   x   x   x   x   x   x   x   x
>   x   x   x   x   x   x   x   x   x
>   x   x   x   x   x   x   x   x   x
>   x   x   x   x   x   x   x   x   x
>   x   x   x   x   x   x   x   x   x
>
> _**Choice 2:**_ Can every number from 1 to 50 be written in more than
> one way as each: a sum, a difference, a product, or a
> quotient? Write your answers as equations. Which was
> easiest: sums, differences, products, or quotients?

CCSS: Operations & Algebraic Thinking: 3.OA, 4.OA   **BIG IDEAS:** 5.1, 5.3
Mathematical Practice: 4

Some students will need the visual stimulus of an array to help them see how
the number 45 can be written in terms of other numbers. Looking at the array,
most students will see sums and products; they can be challenged to look for differences and quotients as well. Rather than specifying multiplication equations
(such as $5 \times 9 = 45$), the question is phrased more broadly, asking simply for equations. This allows an entry point even for students still not comfortable with multiplication (e.g., $9 + 9 + 9 + 9 + 9 = 45$).

Other students will be prepared to work more symbolically. They might select
**Choice 2**, which encourages them to make conjectures and then test them. They may
come to realize that there is an easy way to write every number as a sum ($x + 0$), a
difference ($x - 0$), a product ($1 \cdot x$), or a quotient ($x \div 1$). And they may discover
that coming up with alternative ways to obtain some numbers by writing a product
will not be as easy as coming up with alternatives for the other operations.

In follow-up discussion, all students could be asked what their equations were
and how they arrived at them.

> _**Choice 1:**_ Create two patterns where terms increase by a constant
> amount and where any term in Pattern B is twice as big as
> the corresponding term in Pattern A. Graph each pattern.
> How are the graphs similar? How are they different?
>
> _**Choice 2:**_ Create two patterns where terms increase by a constant
> amount and where any term in Pattern B is a little more
> than three times as big as the corresponding term in
> Pattern A. Graph each pattern. How are the graphs
> similar? How are they different?

CCSS: Operations & Algebraic Thinking: 5.OA   **BIG IDEAS:** 5.8, 5.9
Mathematical Practice: 7

Each of the choices encourages students both to relate two patterns and to
graph patterns.

Parallel Tasks for Grades 3–5

In one case, the relationship is likely easier (e.g., create any pattern and double all terms). The other choice requires a little more thought. A possibility for **Choice 2** might be, for example, the patterns 4, 6, 8, 10, . . . and 13, 19, 25, 31, ... In both cases, the two lines have different slopes.

Questions suiting both groups of students include:

- *Do the corresponding terms of the patterns stay the same distance apart or not? Why or why not?*
- *When you graph the patterns, what "shapes" do you see? Why do those shapes make sense?*
- *Could you have started the first pattern somewhere else? Could you have added a different amount each time?*

SUMMING UP

MY OWN QUESTIONS AND TASKS

Lesson Goal: Grade Level: _____

Standard(s) Addressed:

Underlying Big Idea(s):

_____

Open Question(s):

_____

Parallel Tasks:

Choice 1:

Choice 2:

_____

Principles to Keep in Mind:

- All open questions must allow for correct responses at a variety of levels.
- Parallel tasks need to be created with variations that allow struggling students to be successful and proficient students to be challenged.
- Questions and tasks should be constructed in such a way that will allow all students to participate together in follow-up discussions.

The nine big ideas that underpin work in operations and algebraic thinking were explored in this chapter through nearly 50 examples of open questions and parallel tasks, as well as variations of them. The instructional examples provided were designed to support differentiated instruction for students in both the Pre-kindergarten–Grade 2 and Grades 3–5 grade bands.

Understanding the operations and the properties associated with them as well as understanding pattern are fundamental to work in the elementary grades and beyond.

The examples presented in this chapter are only a sampling of the many possible questions and tasks that can be used to differentiate instruction in operations and algebraic thinking. Other questions and tasks can be created by, for example, using alternate operations or alternate patterns. A form such as the one shown here can be a convenient template for creating your own open questions and parallel tasks. Appendix B includes a full-size blank form and tips for using it to design your own teaching materials.

Expressions & Equations and Functions

DIFFERENTIATED LEARNING activities focused on **algebraic expressions**, equations, and **functions** are derived from applying the Common Core Standards for Mathematical Practice to the content goals that appear in the Expressions & Equations and Functions domains for Grades 6–8.

TOPICS

Before differentiating instruction in work with expressions, equations, and functions, it is useful for a teacher to have a sense of how algebraic topics develop over the grades in the curriculum.

Prekindergarten–Grade 2

Within this grade band, students work with number equations in the context of operations. Most of the equations involve addition and subtraction. Students learn the meaning of an equation as a statement of balance..

Grades 3–5

Within this grade band, students work with number equations in the context of operations. All four operations are involved. They also work with patterns, primarily **linear growing patterns**. They continue to work with the notion that an equation describes a balance.

Grades 6–8

In the middle grades there is a significant amount of work using algebraic expressions, as well as equations. Students use expressions, equations, and formulas to model numerical and real-life situations. They evaluate expressions involving variables and use variables more regularly, recognizing that two different expressions might be equivalent. They solve simple equations and use tables of values to uncover relationships and solve problems. They analyze and solve pairs of **simultaneous linear equations** and relate **linear equations** to various problem situations. They explore, more generally, the concept of function, mostly linear functions.

THE BIG IDEAS FOR EXPRESSIONS & EQUATIONS
AND FOR FUNCTIONS

In order to differentiate instruction in the content areas of expressions, equations, and functions, it is important to have a sense of the bigger ideas that students need to learn. A focus on these bigger ideas, rather than on very tight standards, allows for better differentiation.

It is possible to structure all learning in the topics covered in this chapter around these big ideas, or essential understandings:

6.1. An equation is a statement of balance. The two sides are intended to represent the same value.

6.2. Variables can be used to "efficiently" describe relationships or unknowns.

6.3. The same algebraic expression or equation can be related to different real-world situations or vice versa.

6.4. Operations hold the same meaning when applied to variables as they do when applied to numbers.

6.5. Solving an equation uses relationships between numbers and between operations to determine an equivalent, simpler form of the equation.

The tasks set out and the questions asked about them while teaching expressions, equations, and functions should be developed to reinforce the big ideas described above. The following sections present numerous examples of application of open questions and parallel tasks in development of differentiated instruction in these big ideas across the Grades 6–8 grade band.

OPEN QUESTIONS FOR GRADES 6–8

OPEN QUESTIONS are broad-based questions that invite meaningful responses from students at many developmental levels.

An **inequality** involving the variable m is true when $m = 4$, but false when m = 8. List some possible inequalities.

CCSS: Expressions & Equations: 6.EE, 7.EE
Mathematical Practices: 3, 6

BIG IDEA: 6.1

While some students might choose a very simple inequality, such as $m < 5$, stronger students should be encouraged to think of more complex inequalities. One possibility is, for example, $3m + 5 < 18$. Or some students might be encouraged to use a "greater than" inequality, such as $-3m - 4 > -20$.

> ***Variations.*** A very simple variation is achieved by changing the values for which the inequality is or is not true.

> Make up two equations that use variables and that are true all of the time. Then make up another two equations that use variables and that are true only some of the time.

CCSS: Expressions & Equations: 6.EE, 7.EE   **BIG IDEA:** 6.1
      Mathematical Practices: 3, 7

It is very important that students learn the different ways equations are used. Sometimes equations are used to describe a very specific relationship: for example, $5 + n = 10$ only when $n = 5$. At other times, equations are used to describe broader relationships: for example, $n + 5 = 5 + n$ is true for all values of n and describes a basic property of numbers.

The open task suggested above allows multiple entry points. As an equation that is always true, a student might write a very simple equation, such as $n = n$ or $2 \times n = n + n$, or something more complex, for example, $(2 \times n) + (3 \times n) = 5 \times n$.

Some students might use rational **coefficients** and write equations like $\frac{2}{3}n = 2n \div 3$.

TEACHING TIP. Allowing students to invent all of the values for a question empowers them and truly opens up opportunities.

> Create an equation involving integer **exponents** that you know is true. Tell why it is true.

CCSS: Expressions & Equations: 8.EE   **BIG IDEA:** 6.1
      Mathematical Practices: 6, 8

By not specifying whether the integers need to be positive or negative, this question allows for greater differentiation. While some students will write simple equations, for example, $2^3 \times 2^5 = 2^8$, others might write equations like $2^3 \times 2^{-3} = 1$ or $2^4 \div 4^{-2} = 2^0$ or even $2^3 - 7 = 1$.

TEACHING TIP. Often math teachers who are worried about the standards for their grade insist on specific types of numbers; this makes differentiation more difficult. Being less specific allows entry for all and does not preclude those who are ready for more to actually do more.

> Create an equation involving one variable that has no solutions.

CCSS: Expressions & Equations: 8.EE   **BIG IDEA:** 6.1
Mathematical Practice: 6

Students have a great deal of latitude in creating such an equation; the only requirement is that there is a variable. Some students might use a very simple equation such as $x + 1 = x + 2$, which has no solution. Some students are likely to write an equation such as $2x = 3x$, which, of course, does have a solution, $x = 0$; this error provides the teacher with useful information for assessment of student learning.

Some students might use much more complicated equations, for example, $-3x + 17 = 22x + 10 - 25x$.

> On a 100 chart, you color squares to form a capital letter. If you add the numbers the letter covers, the sum is between 200 and 240. What could the letter be? Where could it be?

1	2	3	4	5	6	7	8	9	10
11	12	13	14	15	16	14	18	19	20
21	22	23	24	25	26	27	28	29	30
31	32	33	34	35	36	37	38	39	40
41	42	43	44	45	46	47	48	49	50
51	52	53	54	55	56	57	58	59	60
61	62	63	64	65	66	67	68	69	70
71	72	73	74	75	76	77	78	79	80
81	82	83	84	85	86	87	88	89	90
91	92	93	94	95	96	97	98	99	100

CCSS: Expressions & Equations: 6.EE, 7.EE   **BIG IDEA:** 6.2
Mathematical Practices: 1, 2, 4, 5

Students can use a simple letter such as an **I** or a more complex one such as a **T**. Some students will solve the question algebraically: for example, the values within an **I** that is four spaces tall are n, $n + 10$, $n + 20$, and $n + 30$. Other students will use a more numerical technique.

As students discuss their solutions, those who did not use variables will have an opportunity to learn from those who describe how they used variables.

> *Variations.* Instead of a letter on the grid, some other type of graphic could be proposed.

TEACHING TIP. Some students might work on this problem numerically and not algebraically. This might be appropriate for students at a certain developmental level, but if students can address the question algebraically, teachers should encourage them to do so.

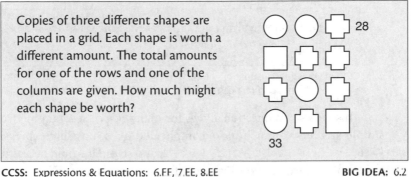

Copies of three different shapes are placed in a grid. Each shape is worth a different amount. The total amounts for one of the rows and one of the columns are given. How much might each shape be worth?

CCSS: Expressions & Equations: 6.FF, 7.EE, 8.EE   **BIG IDEA:** 6.2
Mathematical Practices: 1, 3, 4, 7

Comparing the first row to the first column should help a student figure out why a square must be worth 5, but there is some flexibility in what the other values can be worth. For example, the circle could be worth 10 and the cross worth 8, or the circle could be worth 6 and the cross worth 16. The question can be addressed by using a system of linear equations or more informally.

If all of the row and column values were given, there would be only one solution and the question would be less open. As it is, there are some constraints but some flexibility as well, making the problem open to a broader range of students.

> *Variations.* Information about other rows and columns might have been given. For example, rather than indicating the sum of the first column items as 33, the total for the second column could have been indicated to be 34 or even 40.

There are two unknown numbers, x and y, that are both positive. The only thing you know about them now is that $2x + y = 9$.

Tell some other things about them:

- something you are sure is true about them
- something you are sure is not true about them
- something you are not sure about

CCSS: Expressions & Equations: 6.EE, 7.EE, 8.EE   **BIG IDEA:** 6.2
Mathematical Practices: 1, 3, 7

This is a soft introduction into the uses of simultaneous linear equations. Students will realize that if you have a relationship between two variables like the one given, the values are not yet tied down, but there are things you can conclude.

For example, a student knows for sure that $2x + y$ is not 10 or that $2x + y + 1 = 10$. He or she also knows for sure that if x is a whole number, the value of y must be odd. What a student can't know are the particular values for x and y or which one is greater and which is less.

Students could be encouraged to graph the relationship to help them decide what they do or don't know.

> An expression involving the variable k has the value 10 when $k = 4$.
> What could the expression be?

CCSS: Expressions & Equations: 6.EE, 7.EE   **BIG IDEA:** 6.2
           Mathematical Practices: 3, 5, 7

This very open question allows for many responses, ranging from extremely simple answers to much more complicated ones. Possible expressions include: $k + 6$, $2k + 2$, $14 - k$, $10k \div 4$, $2k^2 - 5k - 2$, and so on. It would be useful for students to identify the expressions that are actually equivalent (e.g., $k + 6$ and $2 + k + 4$) as opposed to the expressions that take on the same values only sometimes (e.g., $k + 6$ and $14 - k$).

➤ *Variations.* This question can easily be adapted by changing either the input or the output value, or both.

TEACHING TIP. Students are often asked to evaluate an expression for a given value of a variable. Questions in which the student is asked to create the expression for a given input/output combination are much more suitable for differentiating instruction.

> A line goes through the origin. Two points on the line are (4,5) and (8,10). Suppose the point (x,y) is on the line. What do you know about the values of x and y?

CCSS: Expressions & Equations: 8.EE   **BIG IDEA:** 6.2
           Mathematical Practices: 3, 4, 7

Students may respond to this question in different ways. While some students might recognize that because the line passes through the origin and has a slope of $\frac{5}{4}$, x is always $\frac{4}{5}$ of y; others might notice that the x value increased by 4 and the y value by 5 and continue that pattern. Still other students might simply suggest that the x is less than the y (which is actually not true when they are negative, but is true when they are positive).

Questions that might be asked include:

- *Do you think that (16,10) is on the line? Why or why not?*
- *What do you think the y-coordinate is when* x *is 3? Why?*
- *Suppose* x *increases by 20, what do you think y increases by? Why?*

If $s = 4$ and $t = 5$, these statements are true:

$$3(s + t) = 27 \qquad 2s + 3t = 23 \qquad 2t - 2s = 2$$

Choose values for p and q. Write three true statements using those variables. See if a partner can figure out what your values are.

CCSS: Expressions & Equations: 6.EE, 7.EE   **BIG IDEA:** 6.2
        Mathematical Practices: 5, 6

Because students can choose the values for p and q and create the equations, even struggling students will be able to participate. For example, a struggling student might choose values such as $p = 1$ and $q = 2$ and use equations such as $p + q = 3$, $q - p = 1$, and $p - 1 = 0$.

When the partner students are given the set of statements created by the first students and asked to figure out the values, the partners need to decide which statement might be the best one to start with. For example, in the set of possible statements from the struggling student given in the preceding paragraph, it is easiest to start with the last equation because it involves only one variable.

Draw a graph of any line. Write an algebraic equation for a different line. Tell which grows faster and how you know.

CCSS: Functions: 8.F   **BIG IDEA:** 6.2
        Mathematical Practices: 4, 5

By allowing students to choose the graph and the equation, the task becomes more comfortable for students without sacrificing any important learning. Whatever graph they choose, they need to figure out its slope, or rate of change. And whatever equation they use, they need to figure out its slope, or rate of change.

It would be interesting to discuss situations where negative slopes are used: *Do we use the absolute value to describe the rate of change or the actual value?*

➤ *Variations.* Rather than a graph or an equation, a student might be asked to use the description of a linear situation that could occur in everyday life.

Describe a function that is not linear by describing what its graph might look like or by describing what the equation might be.

CCSS: Functions: 8.F   **BIG IDEA:** 6.2
        Mathematical Practices: 4, 5

Students have the freedom to describe their non-linear function graphically or algebraically, as well as the freedom to decide what that function is. The algebraic part of this question can be included only if students have been exposed to non-linear relationships, whether functions such as absolute value or **quadratic functions**.

> A certain function is almost linear, but not quite. Create a table of values to show what that function might be.

CCSS: Functions: 8.F   **BIG IDEA:** 6.2
      Mathematical Practices: 4, 5

This question is designed to provide the opportunity for students to discuss what makes a function linear in a variety of ways. While some students will draw a line and wiggle it a bit, using their graph to determine their tables of values, others will recognize that a linear function is one in which the y-values increase or decrease consistently for a fixed change in the x values. They might create tables of values for a linear function and then adjust some of the values.

There might be students who believe that if the y-values are not integers, then the function is not linear. This, of course, is not true, but this question allows for the possibility that this misconception could be dealt with.

> Draw a graph of a function that tells a story. Make sure to use a title and label the axes.

CCSS: Functions: 8.F   **BIG IDEAS:** 6.2, 6.3
      Mathematical Practice: 2

Students should learn that graphs tell stories. A teacher might first show a graph that tells a story and then ask students to create a similar graph. For example, a graph might be labeled "My Bath." The horizontal **axis** could be time and the vertical axis the depth of water in the tub. The graph could show the water level gradually going up as the tub fills, then quickly going up a significant amount as someone enters the tub, and then gradually dropping down. Or a graph might be labeled "Patrons in a Restaurant," with a horizontal axis showing time of day and a vertical axis showing number of patrons at different times.

> Describe three very different situations, all of which could be described by the expression 3x + 2.

CCSS: Expressions & Equations: 6.EE   **BIG IDEA:** 6.3
      Mathematical Practices: 2, 7

Students need to realize that any algebraic expression describes many situations, and it is the structure of the situation that they have in common. In this case, there are either 3 equal groups and 2 more items, or there is some number of groups of 3 and 2 more items. For example, situations such as the following would fit the expression:

- *The cost of 3 identical shirts and a pen that costs $2*
- *The area of a rectangle with a width of 3 joined to a 2 × 1 rectangle*
- *The total income if a student sold many raffle tickets at $3 apiece and got an extra $2 donation*

The equation $4x - 5 = 15$ describes two very different situations. What might those situations be?

CCSS: Expressions & Equations: 6.EE   **BIG IDEA:** 6.3
Mathematical Practices: 2, 4

The expression $4x$ might describe some number of 4s or 4 equal things. This distinction might become the basis to create the two different situations, such as:

- *I bought 4 pairs of socks that cost the same amount. I had a $5 off coupon, and I paid $15 in total. How much was each pair of socks?*
- *I bought some packages of socks that cost $4 each. I had a $5 off coupon, but it still cost $15. How many pairs did I buy?*

Or a student might use the same meaning for $4x$, but apply it in different contexts. So one situation might be:

- *I bought some jars of applesauce. Each cost $4. I got a $5 discount and paid $15 altogether. How many jars did I buy?*
- *I cut some identical 4-inch lengths of wood to build a model boat. I found the pieces I had cut were too long, so I trimmed an equal amount from each one. My trimmings totaled 5 inches in length, and my remaining pieces totaled 15 inches in length. How many lengths did I originally cut?*

Create an algebraic expression that you know is:

- Always more than $2m + 1$
- Always less than $2m + 1$
- Sometimes more than $-2m$
- Sometimes less than $-2m$

CCSS: Expressions & Equations: 6.EE, 7.EE   **BIG IDEA:** 6.4
Mathematical Practice: 7

Because operations with variables work in the same way as they do with numbers, students should realize that adding any positive number to an expression always increases it and adding a negative number always decreases it, but that multiplying or dividing has a different effect on positive numbers than it does on negative numbers.

Possible solutions, therefore, might be: $2m + 2$ (always more), $2m$ (always less), $-4m$ (sometimes more), and $4m$ (sometimes less).

➤ *Variations.* Instead of using the given expressions, more complex expressions might be used. As well, the phrase "never more" or "never less" can be used.

> Describe a situation that might lead you to divide the expression $4a$ by the expression $2a$.

CCSS: Expressions & Equations: 6.EE, 7.EE   **BIG IDEA:** 6.4
Mathematical Practices: 2, 4

Because division always involves creating equal groups, whether by sharing, measuring, or comparing, students might use any of these meanings to create situations.

For example, a student might think that the goal is to figure out how many groups of 2 can be created out of a given number of groups of 4. The answer is clearly twice as many groups of 2 as there are groups of 4.

But another student might set up this situation: *If a rectangle has an area of 4a and a length of 2a, what is its width?*

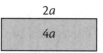

> You multiply two **powers** and the result is 4^{-3}. What two powers might you have multiplied?

CCSS: Expressions & Equations: 8.EE   **BIG IDEA:** 6.4
Mathematical Practices: 3, 8

As stated, the question seems to require students to use a negative exponent. But if a student knows that 4^{-3} is $\frac{1}{64}$, then that student could use positive exponents with fractions of the form $\frac{1}{2}$, $\frac{1}{4}$, $\frac{1}{8}$, and so on.

If students realize they could multiply by 1, they have the option of using a power with a zero exponent. That solution could be, for example, $4^0 \times 4^{-3}$.

> You solve an equation involving negative numbers and/or fractions by first dividing and then subtracting. What might the equation have been? Why would that process make sense?

CCSS: Expressions & Equations: 6.EE, 7.EE, 8.EE   **BIG IDEA:** 6.5
Mathematical Practice: 7

Many students (and teachers) might think that you never divide first and then subtract (thinking of **BEDMAS**), but, in fact, you might well divide first if the equation were $-4(x + 5) = 18$ or even if the equation were $-4x - 20 = 18$. Students might realize that there is always a choice in which operation to perform first, but one operation might seem more efficient to do first in certain situations.

> *Variations.* Instead of dividing and then subtracting, other combinations of operations could be suggested instead.

> The cube root of A is not too much more than the square root of B.
> What could the values of A and B be?

CCSS: Expressions & Equations: 8.EE   **BIG IDEA:** 6.5
          Mathematical Practice: 6

Instead of just asking students to estimate and calculate square roots and cube roots, it seems more interesting to have them relate the two. The required values might be smaller numbers or larger ones. Whereas one student might choose the cube root of 8 (2) to be not much more than the square root of 1 (1), another might choose the cube root of 15,625 (25) to be not much more than the square root of 500 (about 22).

Some students might enjoy the challenge of working with fractions. They could decide that the cube root of almost any fraction close to 1 is not too much more than the square root of any fraction less than 1. For example, the cube root of $\frac{9}{10}$, which is about 0.97, is not that much more than the square root of $\frac{7}{10}$, which is about 0.83.

> The population of one country is about 50 times the population of
> another. Write both populations in scientific notation.

CCSS: Expressions & Equations: 8.EE   **BIG IDEA:** 6.5
          Mathematical Practices: 2, 4

This question should get students looking up information about populations of various countries around the world and finding relationships. They might, for example, find a country with a population of about 50 million (e.g., South Korea) and compare it to a country with a population of about 1 million (e.g., Cyprus). Or they might choose a country like the United States, with a population of about 325 million, and a country with a population of about 6.5 million, such as the Central African Republic.

Then students need to write both of their population numbers in scientific notation. Ideally, they will notice that when they compare the numbers in scientific notation, the values of the exponents for the power of 10 are either 1 or 2 apart, and they will understand why.

> *Variations.* An easier question would be looking for populations where one is a power of 10 times another (e.g., 100 times or 10 times or 1000 times).

> Represent 16^{-8} in different ways.

CCSS: Expressions & Equations: 8.EE   **BIG IDEA:** 6.5
Mathematical Practices: 6, 7

Students can use a variety of exponent properties to come up with alternatives, and there are many. For example, simple choices are 4^{-16} or 2^{-32} or $(\frac{1}{16})^8$. Some students might use operations. For example, expressions could be $2^{-8} \times 8^{-8}$ or $32^{-8} \div 2^{-8}$ or $16^{-8} + 1 - 1$.

The teacher might wish to restrict the alternative representations to be single powers or products of powers.

> The price of 2 coffees and a muffin is \$0.89 more than the price of 2 muffins and a coffee. How much could the coffee cost? The muffin?

CCSS: Expressions & Equations: 8.EE   **BIG IDEA:** 6.5
Mathematical Practices: 2, 4

This is an example of a situation that leads to two equations in two unknowns. The first equation would be of the form $2c + m = a$ and the second equation of the form $c + 2m = b$. Subtracting these two equations results in the equation $c - m = a - b$, or c – m = 89¢ according to the terms of the problem. So any values where the price of the coffee is 89¢ more than the price of the muffin would work. Students still have to choose prices for coffee and muffins that make some sort of sense.

> Two linear equations have a common solution of $p = 3$ and $q = 4.5$. What might the equations have been? What situations might they describe?

CCSS: Expressions & Equations: 8.EE   **BIG IDEA:** 6.5
Mathematical Practices: 4, 5, 7

Students might choose to solve this problem graphically or might use algebraic techniques. They could either draw two lines on the **Cartesian coordinate grid** that cross at (3,4.5) or simply write two equations that correctly describe information about p's and q's. For example, one equation could be $p + q = 7.5$ and another $q - p = 1.5$, or the two equations might be $2p + 4q = 24$ and $p - 2q = -6$.

> **Variations.** The values of p and q can easily be changed.

PARALLEL TASKS FOR GRADES 6–8

PARALLEL TASKS are sets of two or more related tasks that explore the same big idea but are designed to suit the needs of students at different developmental levels. The tasks are similar enough in context that all students can participate fully in a single follow-up discussion.

**Choice 1:** An equation is true only when $x = 5$. What might the equation be? List several possibilities.

**Choice 2:** An equation is true for all values of x. What might the equation be? List several possibilities.

CCSS: Expressions & Equations: 6.EE, 7.EE   **BIG IDEA:** 6.1
Mathematical Practices: 2, 3

One of the important ideas for students at this level to learn is that sometimes we use variables to represent unknown values that make particular equations true. But at other times we use variables to describe relationships or to state "truisms" or identities. For example, the equation $2x = x + x$ is always true; it is an identity. The equation $2x = 4$ is true only when $x = 2$.

Students also learn that many equations can have the same solution, which is dealt with particularly in _**Choice 1**_.

Possible solutions for _**Choice 1**_ include $2x = 10$ and $3x - 7 = 8$. Possible solutions for _**Choice 2**_ are, for example, $3x = 2x + x$ or $5x + 2x = 7x$.

Questions suitable for all students, no matter which task they choose, include:

- _Could your equation have more than one variable in it?_
- _Once you have one equation, how can you get another one from it?_
- _What was the first equation you thought of?_

A company charges a $5 flat fee plus $3 per window to wash windows.

**Choice 1:** How much more would someone pay to have 35 windows washed than 24 windows?

**Choice 2:** Might someone have to pay exactly $87 to have their windows washed? Explain.

CCSS: Expressions & Equations: 6.EE, 7.EE   **BIG IDEA:** 6.2
Mathematical Practices: 1, 2, 4

Both of these questions require students to use the relationship described in the problem set-up. In _**Choice 1**_, the student simply applies the rule to determine the output (price) for a given input (number of windows). This is somewhat more straightforward than **_Choice 2_**, where the student uses either reasoning or algebra to decide whether the suggested final cost is possible. Students with good number sense may quickly realize that because 87 is a multiple of 3, it cannot be 5 more than a multiple of 3 and thus cannot be the total window washing cost.

Choice 1: Choose a location on the 100 grid. Describe how you could get to the numbers that are:

31 less   37 more   18 more

Choice 2: You are on a square called s on the 100 grid. Describe these locations in terms of s:

31 less than s   37 more than s   18 more than s

1	2	3	4	5	6	7	8	9	10
11	12	13	14	15	16	14	18	19	20
21	22	23	24	25	26	27	28	29	30
31	32	33	34	35	36	37	38	39	40
41	42	43	44	45	46	47	48	49	50
51	52	53	54	55	56	57	58	59	60
61	62	63	64	65	66	67	68	69	70
71	72	73	74	75	76	77	78	79	80
81	82	83	84	85	86	87	88	89	90
91	92	93	94	95	96	97	98	99	100

CCSS: Expressions & Equations: 6.EE   **BIG IDEA:** 6.2
Mathematical Practices: 4, 5

These two tasks require the same thinking, although one requires the use of a variable and the other does not. A common discussion will allow students who struggle with the use of variables to gain more insight by listening to those who are using similar reasoning but with variables.

Questions a teacher could ask both groups include:

- _How does the number change if you go to the position right above you? To the position right below you?_
- _How does the number change if you go one position to the left? One to the right?_
- _How do you get to the number that is 31 less? 37 more? 18 more?_

➤ **_Variations._** The values of how much less or how much more can easily be altered.

> **_Choice 1:_** A line passes through (4,3) and (6,–2). What is its equation?
>
> **_Choice 2:_** A line with a VERY steep slope passes through the point (2,–1). What might its equation be?

CCSS: Functions: 8.F   **BIG IDEA:** 6.2
Mathematical Practices: 1, 5

Choice 1 is more likely to appeal to students who prefer as little ambiguity as possible. They might wish to avoid dealing with what a term like "VERY steep" means. But other students might enjoy that challenge and could enjoy the idea of working with a line that is "VERY steep."

Either question provides a teacher with information about whether students can create a line given certain conditions about that line.

> **_Choice 1:_** On a geoboard, create shapes with the indicated number of pegs on the border and inside of them and calculate their areas. Graph the variable that changes against the area. What do you notice?
>
> | Border | Inside | Area |
> |--------|--------|------|
> | 3 | 0 | |
> | 3 | 1 | |
> | 3 | 2 | |
> | 3 | 3 | |
>
> | Border | Inside | Area |
> |--------|--------|------|
> | 3 | 1 | |
> | 4 | 1 | |
> | 5 | 1 | |
> | 6 | 1 | |
>
> **_Choice 2:_** On a geoboard, create triangles with the indicated base and height lengths. Graph the variable that changes against the area. What do you notice?
>
> | Base | Height | Area |
> |------|--------|------|
> | 4 | 1 | |
> | 4 | 2 | |
> | 4 | 3 | |
> | 4 | 4 | |
>
> | Base | Height | Area |
> |------|--------|------|
> | 4 | 2 | |
> | 5 | 2 | |
> | 6 | 2 | |
> | 7 | 2 | |

CCSS: Functions: 8.F   **BIG IDEA:** 6.2
Mathematical Practices: 1, 4, 5

Some students will find **_Choice 2_** accessible because it is more obvious how to create the required shapes and easier to determine their areas.

This task is included in this chapter rather than in the measurement chapter because it is the functional nature of the task that is the emphasis. Students will see that, in each case, there is a linear relationship because one variable in the geometric situation is held constant. For **_Choice 1_**, the function rule that relates the number

Parallel Tasks for Grades 6–8

of inside pegs to the area is $A = I + \frac{1}{2}$; the function rule that relates the number of border pegs to the area is $A = \frac{B}{2}$. For **_Choice 2_**, the function rule that relates the height to the area is $A = 2h$; the function rule that relates the base to the area is $A = b$. When students graph the relationships, each time they will have a straight line whose slope they could interpret.

Questions that suit both tasks include:

- *How are your two graphs alike? How are they different?*
- *What does the slope of the graph tell you?*
- *Why does this particular slope make sense to you?*

Choice 1: List several equations that are true about rectangles with length = 6 units and width = 4 units.

Choice 2: List several equations that are true if T-shirts cost $10 and pants cost $12.

CCSS: Expressions & Equations: 6.EE, 7.EE   **BIG IDEA:** 6.3
Mathematical Practices: 2, 4

Any situation that can be described algebraically can probably be described algebraically in many ways. In this case, **_Choice 1_** allows students to consider how knowing the length and width of a rectangle provides lots of other information, for example, $P = 2(6 + 4)$ (**perimeter**) or $A = 6 \times 4$ (area), or just $l - w = 2$ (comparing length to width). **_Choice 2_** allows students to alter the numbers of each of the items one might buy in order to create different equations, for example, $P = T + 2$ or $12T = 10P$.

Questions suitable for students working with either task include:

- *Would adding the two given values be useful information? Why or why not? What equation would it lead to?*
- *Would doubling the two given values and then adding them be useful information? Why or why not? What equation would it lead to?*
- *Would subtracting the two given values be useful information? Why or why not? What equation would it lead to?*

Choice 1: What situation might be described by the equation $2x + 3y = 100$?

Choice 2: What situation might be described by the equation $3x + 6y = 10$?

CCSS: Expressions & Equations: 6.EE, 7.EE   **BIG IDEA:** 6.3
Mathematical Practices: 2, 4

The difference between these two choices is subtle. In **_Choice 1_**, there are many integer solutions (e.g., $y = 2$ and $x = 47$, or $y = 10$ and $x = 35$) and so there are many simple situations that could be offered. For example, *There were some bikes and some tricycles at a park. There were 100 wheels altogether. How many bikes and how many tricycles?* But in **Choice 2**, even though the numbers might initially appear more comfortable, the solutions cannot be whole numbers, so the situation cannot involve something that can be counted (like bicycles and tricycles). An example for the second situation might be: *Three identical packages of almonds and six identical packages of hazelnuts had a mass of 10 kg. How many kilograms of almonds and how many kilograms of hazelnuts might there have been in each package?* (e.g., 1 kg of hazelnuts in each package and 1.333 kg of almonds in each package).

Order these values from least to greatest. Will your order be the same no matter what the value of n is? Explain.

Choice 1: $\frac{n}{2}$ $3n$ n^2 $3n + 1$ $10 - n$

Choice 2: $4n$ $3n$ $10n$ $3n + 1$ $5n + 2$ $-n$

CCSS: Expressions & Equations: 6.EE, 7.EE
Mathematical Practices: 2, 3, 6   **BIG IDEA:** 6.4

Students need to understand the role of a variable in describing a relationship. Some will consider only positive whole numbers and may need prompting to consider negative numbers or fractions. They can use a variety of strategies to come to a conclusion. For example, students might use a combination of reasoning, visual representations, and **substitution** to order their values. **Choice 1** is more challenging because there is more variation in the type of relationships used.

The following questions are suitable for students who worked on either task:

- *Suppose* $n = 0$. *Would your ordering change?*
- *Suppose* $n = -1$. *Would your ordering change?*
- *How do you know that* $3n < 3n + 1$ *no matter what value* n *has?*

Choice 1: An equation with integer coefficients has a rational number solution. What could the equation be?

Choice 2: An equation with rational coefficients has an integer solution. What could the equation be?

CCSS: Expressions & Equations: 8.EE
Mathematical Practices: 1, 7   **BIG IDEA:** 6.5

In both instances, students are working with equations and rational numbers. The first task might be simpler for struggling students since an easy equation such as $3x = 5$ might work, or $3x = -5$. The second task might take more thinking. Possibilities include any equation of the form $\frac{a}{b} \times = \frac{c}{b}$, where c is a multiple of a; an example might be $-\frac{3}{5}x = -\frac{33}{5}$; more complex equations are also possible in either situation, allowing for further differentiation.

Questions that suit both groups of students include:

- *How many variables will you use in your equation? Why?*
- *Is an integer (e.g., −3) a rational number?*
- *Can an equation with all possible positive coefficients have a negative solution or not? Explain.*

Choice 1:   One gym membership plan is $20 a month and an additional $4 an hour for each hour used. Another is $50 a month and an additional $2 an hour for each hour used. Which membership plan is better? Under what circumstances?

Choice 2:   One membership plan is better than another if you spend fewer than 15 hours per month at the gym, but worse if you spend more than 15 hours per month. What could the two membership plans be?

CCSS: Expressions & Equations: 8.EE   **BIG IDEA:** 6.5
Mathematical Practices: 2, 4, 5

While **Choice 1** might appeal to students who like to be given all the necessary information to solve a problem, **Choice 2** might appeal to students who are comfortable working backward. These tasks were designed so that the 15 hours mentioned in **Choice 2** is the solution in **Choice 1**; this will make common questions easier to arrive at.

The student solving **Choice 1** might draw a graph and see that the intersection of the lines $y = 20 + 4x$ and $y = 50 + 2x$ is at (15,80).

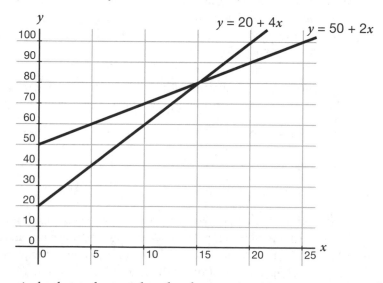

Alternatively, the student might solve the equation $20 + 4x = 50 + 2x$ to determine when the costs would be the same. In either case, they can see that the $20 + $4 per hour plan is better for fewer than 15 hours but worse for more than 15 hours.

A student solving **_Choice 2_** might draw graphs of two lines that intersect at $(15, y)$ for some chosen value of y, for example, $y = 100$. Then he or she would determine the equations of those two lines, for example, $y = 40 + 4x$ and $y = 10 + 6x$. So one plan would have a $40 registration fee and $4 hourly rate and the other would have a $10 registration fee and $6 hourly rate.

Questions that suit both groups include:

- *How could a graph help you solve your problem?*
- *How could algebra help you solve your problem?*
- *Why does it make sense that one plan might be better than another for a lot of hours but worse for fewer hours?*

Choice 1: An equation with one coefficient of $-\frac{2}{3}$ is very easy to solve. What might it be?

Choice 2: An equation with one coefficient of -2 is very easy to solve. What might it be?

CCSS: Expressions & Equations: 6.EE, 7.EE   **BIG IDEA:** 6.5
       Mathematical Practices: 3, 5, 7

Some students will be more comfortable working with non-integer coefficients than others, and that might affect the task they select. In both cases, though, students provide useful information to teachers about their comfort levels in solving equations.

For example, one student might think the equation $-\frac{2}{3}x = -\frac{2}{3}$ is quite easy to solve; another might think $-\frac{2}{3}x = 0$ or $-\frac{2}{3}x - \frac{1}{3}x = -1$ is easy.

Similarly, a student might think the equation $-2x = -2$ is easy to solve, as is $-2x = 0$ or $-2x + 2x = 0$.

Questions suitable for students working with either task include:

- *What do you think makes an equation easy to solve?*
- *Are equations with positive numbers always easier to solve than those with negative numbers or only sometimes?*
- *What properties of operations or numbers can you use in coming up with possible equations?*

Choice 1: Draw a graph that shows how far someone might get on a highway in different amounts of time.

Choice 2: Draw a graph that shows how far someone would get in different numbers of minutes if she was driving 60 mph.

CCSS: Expressions & Equations: 8.EE   **BIG IDEA:** 6.5
       Mathematical Practices: 1, 5

The difference between these two tasks is, to an extent, their specificity. The second task tells students the speed to use, although it does ask about minutes when the rate is given in hours; some decisionmaking has to happen. The first task is more open in that a student is required to decide on a speed, and the units are left open.

So one student might select ***Choice 1***, choose a rate of 60 mph and label one axis in hours and the other in miles; this is fairly straightforward. Another might choose a different rate or might label the axes in different units.

A student who selects ***Choice 2*** has to think about what it means to be able to apply the mph speed for a particular number of minutes. He or she might still create the graph described for ***Choice 1***, recognizing that 1 hour is 60 minutes and just thinking of the time axis as having a **scale** of 60 per unit. Or the student might convert 60 mph to 1 mile per minute and draw a different-looking graph.

Questions that suit both groups of students include:

- *What do you notice about the shape of your graph? Why does that make sense?*
- *What choices did you have in labeling your axes?*
- *How would your graph change if the speed was just a little faster? Just a little slower?*

SUMMING UP

MY OWN QUESTIONS AND TASKS

Lesson Goal: Grade Level: _____

Standard(s) Addressed:

Underlying Big Idea(s):

_____

Open Question(s):

_____

Parallel Tasks:

Choice 1:

Choice 2:

_____

Principles to Keep in Mind:

- All open questions must allow for correct responses at a variety of levels.
- Parallel tasks need to be created with variations that allow struggling students to be successful and proficient students to be challenged.
- Questions and tasks should be constructed in such a way that will allow all students to participate together in follow-up discussions.

The five big ideas that underpin work in the areas of expressions, equations, and functions were explored in this chapter through almost 40 examples of open questions and parallel tasks, as well as variations of them. The instructional examples provided were designed to support differentiated instruction for students in the Grades 6–8 grade band.

The examples presented in this chapter are only a sampling of the many possible questions and tasks that can be used to differentiate instruction in working with expressions, equations, and functions. Other questions and tasks can be created by, for example, using alternate situations or expressions. A form such as the one shown here can be a convenient template for creating your own open questions and parallel tasks. Appendix B includes a full-size blank form and tips for using it to design your own teaching materials.

Measurement & Data

DIFFERENTIATED LEARNING activities focused on measurement and data are derived from applying the Common Core Standards for Mathematical Practice to the content goals appearing in the Measurement & Data domain for Prekindergarten–Grade 2 and Grades 3–5.

TOPICS

Before differentiating instruction in measurement and data, it is useful for a teacher to have a sense of how measurement and data topics develop over the grades.

Prekindergarten–Grade 2

Within this grade band, students compare and measure a variety of **attributes** of objects, including length, weight, time, and money; they work with both **nonstandard** and **standard units**. They tell and write times using hours and half-hours. They also sort objects into categories and answer questions about data sorts. They draw **picture graphs** and **bar graphs** with single-unit scales to represent data. Much of the data they collect relates to measurement.

Grades 3–5

Students tell and write time to the nearest minute and solve time problems. They also measure liquid volumes and masses using metric units, and solve problems about these measurements. They explore perimeter and area of shapes, as well as volume, and become acquainted with some related formulas. They explore units in both the metric and the Imperial systems for length, mass, and time, and they perform unit conversions. They learn about angles as a type of measurement, and they measure angles. They learn to draw scaled picture and bar graphs and generate data through measurement. They create **line plots** based on measurement data.

Grades 6–8

Continued work in the areas of measurement and data for Grades 6–8 is covered in the Common Core domains of Geometry (in the case of measurement) and Statistics & Probability (in the case of data). Therefore, for this grade band, these topics are discussed in Chapters 8 (Geometry) and 9 (Statistics & Probability), rather than in this one.

THE BIG IDEAS FOR MEASUREMENT & DATA

In order to differentiate instruction in measurement and data, it is important to have a sense of the bigger ideas students that need to learn. A focus on these big ideas, rather than on very tight standards, allows for better differentiation.

It is possible to structure all learning in the topics covered in this chapter around these big ideas, or essential understandings:

7.1. A measurement is a comparison of the size of one object with the size of another, often a **benchmark**.

7.2. Accuracy in measurement requires a focus on the attribute being measured.

7.3. The same object can be described by using different measurements.

7.4. The numerical value attached to a measurement is relative to the measurement unit.

7.5. It is sometimes easier to measure an object by breaking it up into pieces.

7.6. The use of standard measurement units simplifies communication about the size of objects.

7.7. Sometimes measurements of a shape are related; knowing one or more measures might reveal the value of others.

7.8. Organizing, representing, and interpreting data are ways to compare objects or situations.

7.9. Graphs are powerful data displays because they quickly reveal a great deal of information.

The tasks set out and the questions asked about them while teaching measurement and data should be developed to reinforce the big ideas listed above. The following sections present numerous examples of application of open questions and parallel tasks in development of differentiated instruction in these big ideas across the Prekindergarten–Grade 2 and Grades 3–5 grade bands.

OPEN QUESTIONS FOR PREKINDERGARTEN–GRADE 2

OPEN QUESTIONS are broad-based questions that invite meaningful responses from students at many developmental levels.

> What can you find in the classroom that is about as long as your arm?

CCSS: Measurement & Data: K.MD, 1.MD   **BIG IDEA:** 7.1
              Mathematical Practices: 5, 6

This question allows the teacher to see what students know about length comparisons. Watching how students compare their arm with other objects shows whether they understand the need for a **baseline** to compare length. Use of the term *about* provides enough latitude that even students who struggle can answer successfully.

➤ *Variations.* This task can be adapted by having students look for objects that are either *shorter than* or *longer than* another item instead of *about the same length*. The item that is used as the basis for comparison can also be varied.

TEACHING TIP. When students are working with length comparisons, it is useful to have spools of string or colored yarn available for them to work with.

> You want to describe to someone how long your pencil is. You can't show the person the pencil, but have to use words. How could you help the person understand its size?

CCSS: Measurement & Data: K.MD, 1.MD, 2.MD   **BIG IDEA:** 7.1
              Mathematical Practices: 3, 5

This question is designed to help students see that the way a measurement is described is invariably in relation to another measurement. If, for example, a student says how many inches long the pencil is, he or she is comparing it to 1 inch. If it is described as being about as long as a hand, the comparison is being made to a hand. In fact, the only way measurements can be described meaningfully is through comparison.

The question is open in that it is completely up to the student to describe the object.

To stress the concept that a comparison is being made, the teacher could explicitly ask: *What measurement comparison are you making when you say that?*

> Describe three things that weigh less than a shoe. Tell how you know
> they weigh less than a shoe.

CCSS: Measurement & Data: K.MD   **BIG IDEA:** 7.1
Mathematical Practices: 3, 5

Asking for items that weigh less than a shoe provides a great deal of latitude for students; they only need to think of very light items. Some students will think of food items, others classroom supplies, others pieces of clothing, and others items such as toys.

In asking how students know the item weighs less than a shoe, it does not matter what items students chose. They will show their understanding or lack of understanding about how to use a balance or scale to compare weights.

➤ *Variations.* Variations of the question can be created by suggesting an item other than a shoe or by allowing students to find heavier items rather than lighter ones.

> You are reading a clock with hands, and the hands are really close
> together, but not on top of each other. What time might it be?

CCSS: Measurement & Data: 2.MD   **BIG IDEA:** 7.1
Mathematical Practices: 5, 6

Students in this grade band learn to read analog clocks to various degrees of precision. Asking the question in this way is likely to encourage students to think of a lot of different possible times, providing a great deal of time-reading practice. The phrase "not on top of each other" is included to ensure that misconceptions about the placement of the hour hand (e.g., thinking the hour hand is precisely on the 6 at 6:30) are not fostered. A student answering 6:30 would be correct, and the teacher could point out how the hands are close, but not on top of each other. It is just as likely a student would say 6:35, which would also be correct. There are many reasonable responses, and many students can be successful.

> Think of some objects in your house that are really long. Tell about
> how long you think they are and why you think you are right.

CCSS: Measurement & Data: K.MD, 1.MD   **BIG IDEA:** 7.1
Mathematical Practices: 1, 3, 6

Using the phrase *really long* generally has the effect of encouraging students to think deeply about the length comparisons they make. The fact that they are thinking about their own possessions makes the question more engaging. To describe the lengths of the selected objects, students have a choice of using comparative language, nonstandard units, or standard units; in this way, in particular, the question becomes open to a broad range of learners.

Likely objects include garden hoses, sofas, beds, driveways, and so forth.

> ➤ *Variations.* Instead of asking for very long items, a teacher could ask for items that hold a lot, items that cover a lot of space, or particularly short items.

TEACHING TIP. Often using an ambiguous term such as *really long* makes a mathematical question much less intimidating to students than using a specific value.

> One length looks shorter than another, but it really isn't. How is that possible?

CCSS: Measurement & Data: K.MD, 1.MD   **BIG IDEA:** 7.2
        Mathematical Practices: 2, 5, 6

An important part of comparing lengths is recognizing that both items must be "straightened out" before they can be compared. For example, a very long coiled rope can look shorter than one that is straight but actually shorter.

Instead of setting out a coiled rope and a straight rope for students to compare, an open question such as the one proposed allows students to imagine all sorts of possible scenarios, providing access to a broader range of students. For example, a student who might have difficulty with the idea of a coiled rope being shorter might be able to imagine an object that is bent around a corner with only some of it visible.

> One container is MUCH taller than another but holds a LOT less. How else might the measurements of the containers be related?

CCSS: Measurement & Data: 1.MD, 2.MD   **BIG IDEA:** 7.2
        Mathematical Practices: 1, 5

Students have a great deal of latitude in choosing dimensions for each container. For example, a student might decide the tall container is 20" tall and the short one 5" tall. But the tall container might be very, very narrow and the short one very, very wide, for example, 1" wide vs. 100" wide.

> Show a BAD way to decide how many pencils long a table is. What makes that way a bad way?

CCSS: Measurement & Data: 1.MD, 2.MD   **BIG IDEA:** 7.2
        Mathematical Practices: 3, 5

There are many details that go into a correct length measurement. These include starting at the beginning and going to the end, leaving no gaps or overlaps, using equal size units throughout, measuring the correct attribute of an item, and staying along the edge of the item.

By asking students to think of a bad way to measure, a teacher gets insight into which of these issues about measurement students really focus on and which they do not.

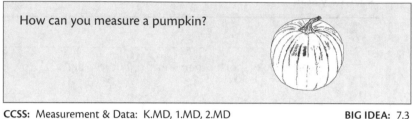

How can you measure a pumpkin?

CCSS: Measurement & Data: K.MD, 1.MD, 2.MD   BIG IDEA: 7.3
       Mathematical Practices: 4, 5

A student can think of many different ways to measure a pumpkin: width, height, weight, the time it takes to grow, or how much pulp it has. Leaving the question open allows each student to choose a measurement on which to concentrate.

Class discussion about the various types of measurements students used will allow them to recognize the fact that there is usually more than one way to measure an object.

➤ *Variations.* Rather than a pumpkin, another interesting object can be selected. For example, students can be asked to measure a bag of popcorn, a car, or a container.

TEACHING TIP. Items to be measured can be chosen based on themes being pursued in other subject areas of instruction, special occasions, or holidays.

An object is big in one way but small in another way. What might the object look like?

CCSS: Measurement & Data: 1.MD   BIG IDEA: 7.3
       Mathematical Practices: 1, 3, 5

This task allows students to consider either 2-D shapes or 3-D objects and offers great flexibility in the types of comparisons that can be made.

Some students might take the condition to mean that the object is big when compared with another object but small when compared with a third object. For example, a pencil is big when compared with a gnat, but small in relation to the length of a hallway. Other students might consider different attributes of the same object. For example, a very tall, very thin cylinder could be considered big in terms of height but small in terms of width. Still other students might consider two different types of measurement. For example, an object could be very heavy (i.e., big in weight) but still very small in width or length or height.

The class discussion should provide the opportunity for the teacher to explicitly make the point that when the size of an object is described, it is important to be very clear about what attribute of the object is being considered.

How many baby steps are there in a giant step?

CCSS: Measurement & Data: 1.MD   **BIG IDEA:** 7.4
        Mathematical Practices: 1, 3, 5

In the earlier grades, students begin to understand how units can be used to describe measurements by working with nonstandard units. Using a unit of a baby step to measure a giant step is meaningful to students. Because of the diversity of possible correct answers, this focus on nonstandard units provides the opportunity for students to later value the usefulness of standard units.

In the short run, this question is easy for any student to act out. It also provides some flexibility, because students are free to interpret baby steps and giant steps in any way they wish.

➤ *Variations.* Variations of the question can be created by using other familiar measures that come in two sizes, for example, *How many baby cups full of water do you need to fill a large glass of water?*

You measure your arm correctly twice, but one time it is 6 units longer than another time. What units might you have been using each time?

CCSS: Measurement & Data: 1.MD   **BIG IDEA:** 7.4
        Mathematical Practices: 3, 5

One way to help students realize that the size of the unit affects the number of units required to measure something is to have them measure the same thing using different units. A more interesting approach, though, might be to tell them the number of units required (or in this case the difference between two measured values) and ask them to come up with the units.

Presenting the problem in this way leaves a great deal of latitude for students to think of interesting units, but also to think about the relative sizes of those units.

➤ *Variations.* The number 6 can be changed to other values, and instead of measuring an arm, some other item might be selected.

You measured something in both inches and centimeters. One of the measurements was 15. What might you have been measuring? What could the other number have been?

CCSS: Measurement & Data: 2.MD   **BIG IDEA:** 7.4
         Mathematical Practices: 3, 5

By not indicating whether the 15 units was 15 inches or 15 centimeters, students must make a choice. One choice might make it easier for them to think of an appropriately sized object than the other choice.

If they choose to make the 15 represent centimeters (perhaps suggesting they were measuring a pencil), the number for inches will be smaller. If they choose to make the 15 be inches (perhaps suggesting they were measuring a ruler; some do come in 15 inch lengths), the number for centimeters would be greater.

The teacher might discuss how knowing whether one number is greater or less than the other automatically indicates whether the unit used for the 15 was inches or centimeters, and why.

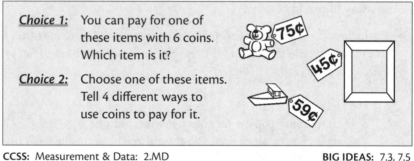

**Choice 1:** You can pay for one of these items with 6 coins. Which item is it?

**Choice 2:** Choose one of these items. Tell 4 different ways to use coins to pay for it.

CCSS: Measurement & Data: 2.MD   **BIG IDEAS:** 7.3, 7.5
         Mathematical Practices: 1, 2, 4, 5

Students who pursue **_Choice 1_** are able to solve a problem involving representing numbers many different ways. They need to recognize, for example, that even though 45¢ can be represented with 9 coins (4 dimes and 5 pennies), it can also be represented with 6 coins (3 dimes and 3 nickels). They must consider all the options for representing each amount.

Those who pursue **_Choice 2_** need only to be able to represent a single number in different ways; with the freedom to choose the number to represent, an even greater likelihood of success for each student is ensured. A student might, for example, choose the 75¢ teddy bear and suggest it could be paid for by using 75 coins (all pennies), 3 coins (3 quarters), 12 coins (7 dimes and 5 pennies), or 5 coins (2 quarters, 2 dimes, and 1 nickel).

Questions to ask all students, no matter which task they chose, would include:

- _What item did you choose? Why did you choose that item?_
- _Could you have used two quarters?_
- _How many coins did you use?_
- _How did you figure out what coins to use?_

> When would it be a good idea to figure out how many units long
> something is by measuring it in pieces? Why would you do it that
> way?

CCSS: Measurement & Data: 1.MD, 2.MD   **BIG IDEA:** 7.5
Mathematical Practices: 1, 3

Different students are likely to come up with different reasons why they might measure a length in pieces. One reason might be that they used a ruler and the item was too long, so they had to measure as much as the ruler length would allow, and then measure the remaining segment of the item. Another situation might occur when an item is not straight and its different parts must be measured separately, for example, a shape like this one:

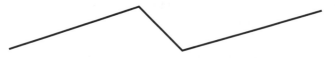

Yet another situation might occur if the student had an orange Cuisenaire rod, which is 10 cm long, as a measuring device; she or he might have to measure several 10s and then some more.

> When might it be a problem if you did not measure in units like
> inches or centimeters, or feet or meters, where we all use exactly the
> same tool size?

CCSS: Measurement & Data: 2.MD   **BIG IDEA:** 7.6
Mathematical Practices: 1, 3

Students might invent scenarios where they are describing to someone else a particular length measurement that needs to be correct (e.g., talking over the phone to Grandpa, who is making an item to replace a broken part of a toy, and its size needs to be exactly right). If the teacher reads students the story *How Big Is a Foot?* by Rolf Myller (Myller, 1991; also available on YouTube), students might get other interesting ideas.

In this question, the openness is in the invention of the situations.

> Draw a shape where if you know one side length of the shape, you
> are sure you know at least one other side length, too.

CCSS: Measurement & Data: 1.MD, 2.MD   **BIG IDEA:** 7.7
Mathematical Practices: 1, 3, 6

In responding to this question, students use geometric knowledge. They might consider squares, rectangles, or perhaps **equilateral triangles** or isosceles triangles even though they might not use that terminology.

> Choose a width for a rectangle. How do the perimeter and area of
> that rectangle change if the width is increased by 2 units?

CCSS: Measurement & Data: 1.MD, 2.MD   **BIG IDEA:** 7.7
       Mathematical Practices: 1, 5, 6

Students have a choice of the original width and this will result in different answers for how the perimeter and area change. For example, if the width changed from 4 cm to 6 cm, the perimeter increases by 4 cm, but the area might increase by any number, depending on the length of the rectangle. If the old rectangle had been 4 cm × 9 cm, the new area would be increased by 18 cm², but if the old rectangle had been 4 cm × 20 cm, the new area would be increased by 40 cm². If the width, though, changed from 8 cm to 10 cm, even though the perimeter still increases by 4 cm, the area, again, can increase by any number of square centimeters, depending on the length.

> [**TO THE TEACHER:** Using sorting materials of any type, choose four
> items that are similar and different in various respects.]
>
> Look at the items the teacher has placed in front of you. Which one
> does not belong? Explain. Think about the items in a different way.
> Try to find a different item that doesn't belong if you think about
> them in this new way.

CCSS: Measurement & Data: 1.MD   **BIG IDEA:** 7.8
       Mathematical Practices: 3, 6

Many students are familiar with the game "Which Doesn't Belong?" What makes this an open question is that there is usually more than one correct response, especially if the items are carefully chosen. For example, using the shapes below, the most obvious response might be that the triangle does not belong because it is the only white shape. Students might also suggest, however, that the circle does not belong because it is the only round shape or that the second square does not belong because it is the only small shape.

> **Variations.** Other sets of items (drawings or concrete items) and other numbers of items can be used to create new questions. Students can be allowed to create their own groups of items and work with partners who will be asked to find the item that does not belong.

TEACHING TIP. The data strand is ripe for differentiation. It is the *processes* of data collection, data display, and data description that are important mathematically and not the specific data themselves. There is often no reason not to let students collect data about something of personal interest to them.

Think about a question you might ask your classmates to which there are three possible answers. Conduct a survey and then graph the results by using either a **concrete graph** or a bar graph.

CCSS: Measurement & Data: 2.MD   BIG IDEA: 7.8
Mathematical Practices: 1, 5

To collect useful data, students must learn to anticipate possible responses. This question requires such planning by the students, yet allows great flexibility. Students might, for example, ask which of three colors other students like most, which of three television shows other students prefer, or which of three places would be their favorite for a vacation. Students will quickly realize that they need to be careful in how they pose their question. For example, a student might want to ask, *Which pet do you have—a dog, a cat, or a bird?* With such a question, the student would have to be prepared for respondents who want to select two or three or no choices, as well as those who name other pets. Some students might be more comfortable rephrasing the question in a less restrictive way and coming up with their own strategies for dealing with nonstandard answers (e.g., adding an "other" bar to the graph).

To accommodate students with a wide range of graphing abilities, a choice is given for using either a concrete graph or a bar graph, depending on what is most comfortable for each individual. Linking cubes can be provided to make creating a concrete graph easier for students who would benefit.

➤ *Variations.* Variations of this task are not difficult to create. A different number of categories can be requested, or students can be asked to collect data from someone other than classmates.

How could you sort a group of toys into two or three groups and make a graph to show how many toys are in the different groups?

CCSS: Measurement & Data: 2.MD   BIG IDEAS: 7.8, 7.9
Mathematical Practice: 3

It is a good idea for the teacher to help get this activity started by suggesting that students name some of their favorite toys and creating a list on the board. In this way, it will be easier for students to come up with ideas for how they might sort toys. Sorting criteria might be whether or not the toys are electronic, whether or not they are board games, whether younger or older children like to play with them, how many people can play with them at one time, and so on.

Students could be encouraged to make a picture graph, a bar graph, or a concrete graph by using cubes or some other object, depending on student familiarity with the different types of graphs. Because students are free to sort the way they wish and to graph the way they wish, this task is accessible to many students. It appeals to those who can use only simple sorting criteria and to those who can employ more complex criteria. It suits those who can create only concrete graphs as well as those who are comfortable with more abstract graphs.

TEACHING TIP. When young students are asked to collect data, it often makes sense to have them work with items in their everyday life. However, there may be times when it is beneficial to shift the focus to mathematical items, such as numbers or shapes, to integrate different mathematics strands.

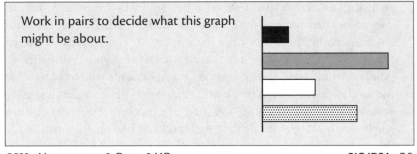

Work in pairs to decide what this graph might be about.

CCSS: Measurement & Data: 2.MD   **BIG IDEA:** 7.9
Mathematical Practice: 3

Because the graph is deliberately vague, offering no title or labels, students have a great deal of latitude in deciding what it might be representing. For example, one pair might suggest that it is a graph showing the favorite kind of pizza for a group of teens. The students should then have to explain what each bar represents (e.g., pizza with olives, pepperoni, or extra cheese, or vegetarian pizza) and why they believe the values make sense (e.g., more people like pepperoni than anything else, very few teens like olives, etc.). The students can also be encouraged to suggest the values that they think each bar might represent. Although, of course, there is no single correct answer, the teacher should steer the discussion, if necessary, to make sure that students' values make sense, for example, that the number for the longest bar is about three or four times the number for the shortest bar.

In the follow-up discussion, the teacher should draw students' attention to the variety of responses about what was represented by the graph. The fact that a graph without labels or a title could reasonably represent so many kinds of information demonstrates what powerful tools graphs can be.

TEACHING TIP. Discussion of a graph without titles and labels provides the teacher with the opportunity to help students understand the difference between a situation where many answers are desirable (e.g., an unlabeled graph) and a situation where everyone is expected to draw the same conclusion (e.g., a graph that is clearly labeled).

Think of a favorite song or poem in which there are not too many words, but some words are repeated. An example is "Itsy Bitsy Spider," where the words *the, itsy, bitsy, spider, spout, rain,* and *and* are among the repeated words. Create a display to show each word in the song or poem and how many times each word appears.

CCSS: Measurement & Data: 1.MD, 2.MD   **BIG IDEA:** 7.9
Mathematical Practices: 5, 6

Allowing students to select their own song or poem is a first step toward differentiating instruction. Further, allowing students to choose their own display method, whether, for example, a **tally chart** or a graph, is another step.

Once the displays have been created, students should have an opportunity to share their work, because doing so provides an opportunity to enhance their mathematical communication skills. For example, a teacher could ask:

- *How did you decide what kind of display to use?*
- *How did you create your display?*
- *How do you know your data are correct?*

> **Variations.** Instead of using a poem or rhyme, students can be asked to simply select a page or a paragraph in a book and perform the same task. To keep the project manageable, students can be asked to choose a limited number of specific words to tally (ignoring the others) or to tally the most common words or the words with a specific number of letters.

TEACHING TIP. When a suggestion is offered in an open question, as was done by mentioning "Itsy Bitsy Spider," students may feel that they must take it. This can narrow the potential for differentiation because students feel less in control of interpreting the question in their own way. Teachers should be judicious about how often this type of scaffolding is provided, recognizing that for some students who have a difficult time getting started, a suggestion may be helpful, whereas for others it can limit expression.

Open Questions for Pre-K–Grade 2

> You are measuring 10 things using inches. When you make a line plot, there are a few tall lines of x's, then a break, and then a few more tall lines of x's. What might you be measuring and what might the lengths be?

CCSS: Measurement & Data: 2.MD   **BIG IDEA:** 7.9
Mathematical Practice: 5

A student should realize that there are a lot of items of similar length and then a lot of other items of a similar length that is greater than the length of the items in the first set. The student has a great deal of choice in what those two sets of similar lengths might be, what the specific items are, and how big the break between the groups might be. For example, a student might measure the arm lengths of some students in second grade and then the arm lengths of some teachers, all to the nearest inch. Or they might measure typical heights of some paperback children's books and some larger textbooks or of small and large water bottles to the nearest inch or the heights of a group of cars and a group of trucks to the nearest foot.

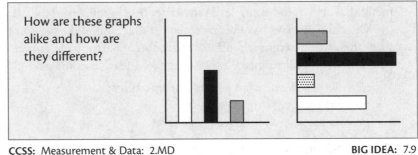

How are these graphs alike and how are they different?

CCSS: Measurement & Data: 2.MD   **BIG IDEA:** 7.9
Mathematical Practices: 3, 6

Asking students to compare two items and notice similarities and differences always supports differentiation. There is deliberately no mention of what the bars represent to extend the potential of the task to meet the needs of a broad range of learners. Many students will focus on the colors of the bars, the number of bars, or the orientation of the bars as either vertical or horizontal. These are useful features to notice, but the teacher should encourage more attention to the underlying mathematics by asking questions such as:

- *How do the categories compare in size?*
- *Does one category have about twice as much as another on either graph?*
- *Can you tell how much more the biggest category is than the smallest one on either graph?*

Open Questions for Pre-K–Grade 2

OPEN QUESTIONS FOR GRADES 3–5

> You want to measure the carpet you will need to fit under your
> hamster's cage. What do you need to measure and how would you
> measure it?

CCSS: Measurement & Data: 3.MD   **BIG IDEA:** 7.1
       Mathematical Practices: 1, 5

The context suggested is one to which most students can easily relate, and it allows for a variety of responses. Whereas some students might think about the amount of space (or area) the cage occupies, others might consider length measures, for example, how long or how wide the cage is.

Students have flexibility because they can imagine the cage to be any size or shape they want and can envision a carpet just fitting the cage or being much longer or wider than the cage. No matter what decision they make, they will still be considering that the cage's size can be described by using a measurement.

> The perimeter of a shape is a little more than three times the length
> of one side of the shape. What could the shape be? Sketch it and
> show its dimensions.

CCSS: Measurement & Data: 3.MD, 4.MD   **BIG IDEAS:** 7.1, 7.7
       Mathematical Practices: 1, 4, 5

There are different strategies students can use as starting points for this question. Some students will begin with a shape—which might be a rectangle, a triangle, or some other shape—and try to make the conditions work. Other students will simply choose a side length, triple it, and then try to make the perimeter come out to be that tripled number.

For example, a student could decide to use a triangle. If two sides are equal and one is only slightly longer than the other two, the conditions are met. Or a student could use a rectangle and make sure the width was just slightly more than half the length (e.g., if $l = 10$ and $w = 6$, the perimeter is 32, just a little more than 3×10).

Alternately, the student could decide one side of a hexagon is 10 and the perimeter is almost 31 and create this shape:

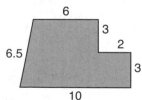

It is primarily because the shape can be simple or more complex that the question is open.

> **Variations.** The problem can be changed by altering the conditions on the perimeter. For example, the perimeter might be required to be slightly less than four times one side length.

> Look at a page of a newspaper that has both advertising and news. Which area is greater—the area for the ads or the area for the news?

CCSS: Measurement & Data: 3.MD, 4.MD   **BIG IDEA:** 7.1
         Mathematical Practices: 1, 4, 5, 6

Area becomes a more important topic in this grade band. As with other measurements, it makes sense to begin the study of area by directly or indirectly comparing measurements rather than using units. In this open question, students have a choice of comparing the measurements in whatever way they wish. Some might cut up the page and overlap the various parts. Others might, independently, come up with the idea of using some sort of unit of measurement.

Follow-up conversation among all the students will allow each one to learn how others approached the problem. In this way, all students are likely to expand their understanding of the concept of area.

> ➤ *Variations.* Rather than comparing parts of a newspaper, students can be asked to compare other areas that might be interesting to them, for example, parts of a board game that are of different colors.

> Make up a measurement problem to solve where you include some measurement information that is not used in the solution. Then solve the problem. Explain why some of the measurements were not needed for solving the problem.

CCSS: Measurement & Data: 4.MD   **BIG IDEA:** 7.2
         Mathematical Practices: 1, 2, 3

It is fairly unusual to ask students to include irrelevant information, but it makes sense to do so when trying to focus them on the notion that in measurement situations, there is often both relevant and irrelevant information.

A student might create a situation such as this one:

My dad drives 20 miles to work each day; he also drives home that distance. He goes to work 5 days a week. Each drive takes him about half an hour. How far does he drive to and from work each week?

The student needs to realize that the time factor is irrelevant.

> The areas of two shapes are almost the same, but the perimeters are very different. What might the shapes be?

CCSS: Measurement & Data: 3.MD, 4.MD   **BIG IDEA:** 7.3
         Mathematical Practices: 1, 3, 7

It is not uncommon to have students compare the areas of two rectangles with the same perimeter or the perimeter of two rectangles with the same area. By not specifying the shapes in the question and not requiring that the shapes be the same, students have much more freedom to work with shapes of their own choice. Additional flexibility is provided by suggesting that the areas are almost the same, rather than exactly the same.

As students create the required shapes, they will likely get a great deal of practice with calculating perimeters and areas. One solution is shown here:

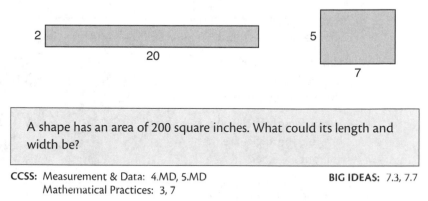

> A shape has an area of 200 square inches. What could its length and width be?

CCSS: Measurement & Data: 4.MD, 5.MD   **BIG IDEAS:** 7.3, 7.7
Mathematical Practices: 3, 7

By not specifying what the shape is, this question is much more open than it would otherwise be. Some students will decide to use a square. They may struggle a bit because 200 is not a perfect square, but they may recognize that the square's length and width would be about 14 inches. Other students will choose a simpler strategy, using a rectangle of length 20 and width 10. Still other students will choose a shape with more sides, particularly if grid paper is provided that would allow them to experiment. Students should be encouraged to determine more than one answer so that they use a variety of formulas or alternate methods to create shapes.

➤ *Variations.* The question can be varied by changing the area required or by adding stipulations to the type of allowable shapes. For example, it could be specified that the shape must have four or more sides or that at least two of its sides must be congruent. Another way to vary the question is to give a volume measurement rather than an area measurement.

TEACHING TIP. By providing the area of a shape and not specifying additional information about the shape, students are free to access area formulas that are familiar to them. During the follow-up discussion, they may be exposed to formulas with which they are less comfortable.

> A container holds about 4 liters. Describe its size in other ways.

CCSS: Measurement & Data: 3.MD   **BIG IDEAS:** 7.3, 7.4, 7.6
Mathematical Practice: 6

Some students will answer the question by relating the 4 liters to familiar, known containers. For example, if a student knows that a pitcher at home holds

1 liter, he or she can imagine a pot that holds 4 times as much. Other students will answer the question by suggesting width, depth, or height measurements that are reasonable for a container of this size.

Discussing the various approaches to the problem will help all students understand how knowing the size of one item can assist in determining the size of other items.

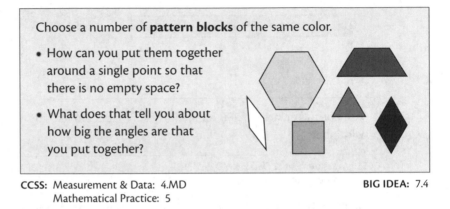

Choose a number of **pattern blocks** of the same color.

- How can you put them together around a single point so that there is no empty space?

- What does that tell you about how big the angles are that you put together?

CCSS: Measurement & Data: 4.MD   **BIG IDEA:** 7.4
        Mathematical Practice: 5

Many times the teaching of angles is introduced with standard units. There is value, however, in thinking about nonstandard units to get across the concept of angle measurement without the additional complication of trying to convey what a degree is. Putting together many copies of the same angle, as is suggested in this task, helps students think about an angle as a nonstandard unit.

Allowing students to choose which shape they use gives them the chance to pick a simple angle (such as a **right angle** in the square) or a more complex angle (e.g., one of the **trapezoid** angles). A connection to degrees can also be made. Students will notice that 3 hexagons can be put together, so the angle measure is $360 \div 3 = 120°$; 6 triangles can be put together, for an angle measure of $360 \div 6 = 60°$; 4 squares can be put together, so the angle measure is $360 \div 4 = 90°$; either 3 or 6 **rhombuses** can be put together, depending on which angles are used; and 12 of the thin **kites** can be put together if the small angles are used, for an angle measure of $360 \div 12 = 30°$.

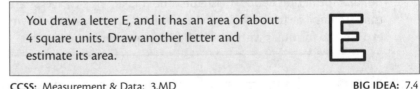

You draw a letter E, and it has an area of about 4 square units. Draw another letter and estimate its area.

CCSS: Measurement & Data: 3.MD   **BIG IDEA:** 7.4
        Mathematical Practice: 5

There are many possible responses students might come up with. A student might choose a much smaller or much larger letter or might try to draw something very close in size to the E. A student might even turn the E sideways and call it a W.

By allowing the student to choose the second letter and its size, there is much less risk to the student of being wrong. Yet the student still applies his or her understanding of what area means.

TEACHING TIP. Differentiated instruction depends on providing some student choice. In this particular task, the student is allowed to choose the letter with which to compare the E.

> What might be units that would make this true? (Each blank is the name of a unit.)
>
> $$6.2 \underline{\hspace{1cm}} = 6200 \underline{\hspace{1cm}}$$

CCSS: Measurement & Data: 5.MD   BIG IDEA: 7.4
Mathematical Practices: 6, 7

Hopefully, students will see that metric units might be useful to solve this problem. Any kilo. . . /base unit combination (e.g., kilometers and meters, kilograms and grams, etc.) would work, as would any base unit/milli. . . combination (e.g., grams and milligrams, meters and millimeters, etc.).

> Describe a situation where you would wish to convert a measurement in millimeters to a measurement in meters.

CCSS: Measurement & Data: 5.MD   BIG IDEA: 7.4
Mathematical Practice: 4

There are a number of different situations a student might come up with. For example, it might be a matter of calculating a perimeter where some side lengths are given in meters and others in millimeters. Or it might be a situation where someone is comparing a measurement given in millimeters to a measurement given in meters, or where someone is asked to figure out the length of a line of pennies, if each penny's width is given in millimeters.

> Draw an angle. What size angle unit (other than 1°) could you use to make your angle measure 6 of those units?

CCSS: Measurement & Data: 4.MD   BIG IDEA: 7.4
Mathematical Practices: 1, 3, 5

Often the only experience in measuring angles that students get is measuring with degrees. But other nonstandard units (wedges of different sizes) could also be used to measure angles. This open question will provide students with the opportunity to explore this idea. They will see that with angles, just as with lengths, areas, masses, and so forth, we need more of a smaller unit and fewer of a larger unit to describe the size of an object. For example, the angle below, to the left, is about six units if the unit is the angle to the right.

> Create a shape that is unusual but whose area you could figure out
> by cutting it up into rectangles.

CCSS: Measurement & Data: 3.MD, 4.MD   **BIG IDEA:** 7.5
Mathematical Practices: 1, 5

Students can apply what they know about the formula for the area of a rectangle, or they might superimpose a transparent grid on top of the shape to determine the area of each part and add the parts together.

The interesting part of the problem comes in the student justifying the areas and ensuring there are no gaps or overlaps.

Students enjoy creating unusual shapes, such as the ones below. The shape could be relatively simple, like the one on the left, or somewhat more complex, like the one on the right.

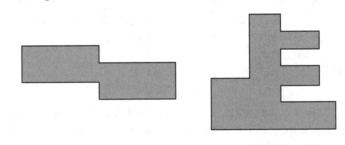

> Choose a time that is between ☐:30 and ☐:45 as the start time for
> an activity. Choose a time that is between ☐:00 and ☐:10 as the end
> time for an activity. Tell how long the activity went on.

CCSS: Measurement & Data: 3.MD   **BIG IDEA:** 7.5
Mathematical Practices: 1, 2, 5

It is highly likely that as students determine the elapsed time from the start to the end of an activity, they will add pieces of hours together to determine the total time. For example, to determine the time between 10:35 a.m. and 12:06 p.m., students are likely to go 1 hour to get to 11:35, another 25 minutes to get to noon, and then 6 more minutes.

Allowing students an opportunity to choose their own start and end times provides useful differentiation.

> ➤ *Variations.* Students might be given a more open choice in selecting start and end times, which could simplify the problem for struggling students.

Two rectangular **prisms** with the same size base are put together to make a larger solid. One of the prisms is VERY short and one is VERY tall. Choose values for the areas of the bases and the two heights and figure out the total volume of the larger shape.

CCSS: Measurement & Data: 5.MD   **BIG IDEA:** 7.5
Mathematical Practice: 5

Students have an opportunity to practice the formula for the volume of a rectangular prism in solving this problem. However, they can choose numbers that are comfortable for them to work with. Suggesting that the heights are very short and very tall will give students the opportunity to have some fun with what these heights might be and allows for practice in either the metric or Imperial system.

Students also get the opportunity to see, again, that we often calculate the measurement of one shape by calculating measurements of parts of that shape and putting those values together.

> ➤ *Variations.* A simple variation is to have different bases but the same height.

A shape is made up of two rectangles. The area of the full shape is 50 square inches. What might the side lengths of the two rectangles be? What might the full shape look like?

CCSS: Measurement & Data: 4.MD   **BIG IDEA:** 7.5
Mathematical Practices: 1, 4, 5

Students have a great deal of latitude in deciding what the two rectangles and the full shape might look like. But no matter what decision is made, most students are likely to use the formula for the area of a rectangle in working through the problem.

For example, a student might decide that the rectangles are equal in area and are both squares, and use two side-by-side squares, as shown at the upper right, each with side length of 5 inches.

But another student might randomly choose a 4" × 6" rectangle for one of the two rectangles, realize that the other rectangle needs an area of 26 square inches, and might then use a 2" × 13" rectangle, as shown at the lower right.

> You build two structures using **snap cubes**. Their volumes are close to the same, but not exactly the same. One is much wider than the other. What could the shapes look like? What could their volumes be?

CCSS: Measurement & Data: 5.MD   **BIG IDEA:** 7.5
        Mathematical Practice: 5

Initially students learn that the volume of an object is described by the number of unit cubes it takes to build it. A question like this one emphasizes that idea.

Students are free to choose the volume they wish, whether great or small. They are free to choose how close to make the two volumes. They are free to decide what "much wider" means. They might choose to build only rectangular prisms and informally or more formally use a formula to determine the volumes, or they could choose to use no formula at all.

One possible solution is to use one structure of volume 10 cubic units and one of volume 11 cubic units. The first structure could be short and wide and the second one tall and narrow.

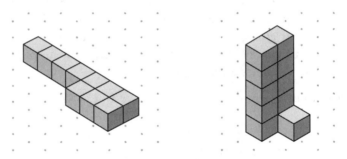

Another student, though, might imagine one prism 1 cube high with a base of 20 cubes by 2 cubes and a second prism 13 cubes high with a base of 1 cube by 3 cubes, using multiplication to help choose those sizes.

➤ *Variations.* Conditions of the problem can be altered. For example, one volume could be exactly twice the other.

> A structure is made up of three separate pieces; each is a rectangular prism. It turns out that if you know the volume of one of those prisms, you automatically know the volume of the other two. How is that possible?

CCSS: Measurement & Data: 5.MD   **BIG IDEA:** 7.7
        Mathematical Practices: 1, 3

One student might decide that all three pieces are identical and thus would suggest that knowing the volume of one of the pieces also tells about the other two. The structure at the left on the next page illustrates one such possibility.

But a different student might decide that two pieces are identical and the third is the same size as putting the first two pieces together, as illustrated at the right below. In this case, as well, only one piece needs to be known to figure out all three. Other solutions are also possible.

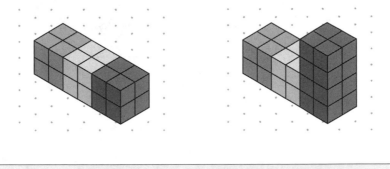

> Choose a width for a rectangle. Make the length a little more than twice as much as the width. Figure out the perimeter. Explain why you didn't need to measure all four sides to figure it out.

CCSS: Measurement & Data: 4.MD   **BIG IDEA:** 7.7
   Mathematical Practices: 1, 2, 4

The problem is made open by allowing students to choose the width as well as what "a little more" might mean. All students, regardless of choice, get practice in determining perimeter. All students, too, explicitly recognize that because opposite sides of rectangles are equal, knowing one length and one width actually yields all four measurements.

> Imagine a pizza order from your school. Students were able to choose among vegetarian, pepperoni, cheese, or meat pizzas. What do you think the bar graph might look like? What scale makes sense to use and why? Pose and solve two problems related to your graph.

CCSS: Measurement & Data: 3.MD   **BIG IDEA:** 7.9
   Mathematical Practices: 4, 5

There are many aspects to this question that make it open. Students can choose the total number of students, choose the scale of the bar graph, and choose the distribution of the data. The topic is also one that is of interest to many students; in fact, the teacher could offer more choice by allowing the students to choose the pizza types.

No matter which choices students make, they will be examining how bar graphs with scales are constructed and will be solving problems related to their graphs. The problems could be simple comparisons (e.g., how many more people chose pepperoni than vegetarian) or they could be multiplicative comparisons (e.g., how many times as many people chose pepperoni than chose vegetarian).

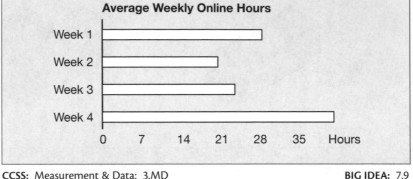

The data below show a student's number of hours online each week over the course of 4 weeks. How would you describe her pattern of online time? If another graph on the same topic looked different, what might it look like?

Average Weekly Online Hours

CCSS: Measurement & Data: 3.MD   **BIG IDEA:** 7.9
Mathematical Practice: 6

This question is open in that how much students describe or what they focus on is purely up to them. One student might focus on how the values increase or decrease over the 4 weeks, another on the fact that the time spent online is considerable for certain weeks, and others on the fact that the values are all greater or less than a certain amount.

In creating a graph that looks different, students have many options. Some students might simply use a different type of graph to display the same data, or they might use a similar graph but with a different scale or a different orientation. Other students might generate data they see as substantively different from the data provided, and then present the new data in a graph similar to the one shown or in a different format.

> ***Variations.*** Different data or a different type of graph can be provided. Students can also be instructed to create a graph that is similar to the given one, rather than different from it.

Select a graph type you would use to display these data. Why is your choice a good way to show the data?

Favorite Dinosaurs	
Tyrannosaurus Rex	25
Triceratops	3
Stegosaurus	8
Brachiosaurus	2

CCSS: Measurement & Data: 3.MD   **BIG IDEA:** 7.9
Mathematical Practice: 5

It is important for students to learn that different types of graphs are appropriate in different situations. By allowing students to choose their own graph type, instead of telling them what graph to use, it is likely that they will think more deeply about the reasons behind using different types of graphs in different situations.

In addition, allowing students to choose the type of graph to use provides opportunities for students with different levels of knowledge to succeed at the task. If, for example, students are asked to create a particular type of graph, those who are less comfortable with that graph type will have less chance for success. With the more open approach, as other students describe their work, even struggling students will gain experience with graph types they might otherwise avoid.

A picture graph has been created, and these two symbols appear somewhere on the graph. Describe as much as you can about what each symbol probably represents. Explain your thinking.

CCSS: Measurement & Data: 3.MD   **BIG IDEA:** 7.9
        Mathematical Practice: 3

Learning to use a scale indicator in a graph is a big step for this grade band. Students need to recognize that the use of partial symbols is often necessary when a scale is used. In this case, students can use visual skills to surmise that the scale is probably a multiple of 4 because a quarter symbol and a half symbol were used. Some students will assume that the scale must be 4; others will realize it could be a multiple of 4. Some more advanced students may realize that there is no way to know. For example, if a whole circle represents 3, the half symbol could represent half that amount and the quarter symbol one fourth of that amount, assuming that the data represent fractions and not whole numbers.

TEACHING TIP. Many students struggle with picture graph symbols involved in scales. For example, rather than using a circle to represent four items, they might use a shape with four marked sections if they are uncomfortable using a single item to represent 4.

A line plot shows the distances children in a group can jump, each distance to the nearest quarter of a yard. Draw a line plot that you think is reasonable. Tell why you think it is reasonable by describing why you used the data you did.

CCSS: Measurement & Data: 4.MD   **BIG IDEA:** 7.9
        Mathematical Practices: 3, 5

By asking the question in this way, students are required to actually experiment to see what distances are reasonable. In addition, they get practice in creating line plots and give some thought to the need for the data to be somewhat variable. They might consider factors such as gender or how athletic the group might be (e.g., is this a group of athletes or just a random group of children?). This kind of thinking is a preview of more deliberate investigation of the notion of variability in the next grade band.

> Create a line plot that might show the measurements of finger lengths to the nearest $\frac{1}{8}$ of an inch. Ask and answer two questions about your plot—one that would require you to multiply and one that would require you to divide.

CCSS: Measurement & Data: 5.MD   **BIG IDEA:** 7.9
Mathematical Practices: 3, 5

Rather than starting with data, this task gives students the opportunity to use a sufficient sample to generate what they would see as reasonable data.

A further level of choice makes for an even more open question—students get to decide what questions to ask about the plot. The requirement that they use certain operations forces them to think about situations that match each operation. For example, division might be used to calculate an average (a **mean**); multiplication might be used if the students speculated on what would happen if there were another measure that had as many pieces of data as one already present in the graph.

> You create a line plot that is based on measuring items to the nearest quarter of an inch. You notice that your plot looks a lot like a steep mountain. What might the measurements be and what might you be measuring? Why does it make sense that your plot would look like a steep mountain?

CCSS: Measurement & Data: 3.MD, 4.MD   **BIG IDEA:** 7.9
Mathematical Practices: 2, 3

A question like this makes the most sense after students have had a number of opportunities to see line plots of various shapes and line plots based on measurements to the nearest quarter of an inch. They then have a sense of how these plots look alike and different and what measurements might be reasonable. The interpretation of a "steep mountain" is flexible, but the interesting part is why students think there might be many items with a single measurement with relatively little variation.

> **Variations.** A different shape might be suggested, rather than a steep mountain, for example, a shape that is almost a rectangle or a very spread-out shape.

Open Questions for Grades 3–5

PARALLEL TASKS FOR PREKINDERGARTEN–GRADE 2

> **PARALLEL TASKS** are sets of two or more related tasks that explore the same big idea but are designed to suit the needs of students at different developmental levels. The tasks are similar enough in context that all students can participate fully in a single follow-up discussion.

Decide which is longer.

Choice 1: The distance from your shoulder to your wrist or the distance around your head

Choice 2: The distance from your elbow to your wrist or the length of your foot

---

CCSS: Measurement & Data: K.MD, 1.MD, 2.MD   **BIG IDEA:** 7.1
Mathematical Practice: 5

These tasks differ in that students can literally move their foot up to their arm to compare lengths for *Choice 2* but need to use an indirect method to compare measurements (e.g., using a string) for *Choice 1*.

Whichever choice was selected, students could be asked:

- *Can you just look to decide which is longer?*
- *What do you have to do to check?*
- *What other length comparisons would be easy to make?*
- *Which would be trickier?*

TEACHING TIP. When numbered choices are offered, the "simpler" option should sometimes be presented as Choice 1 and other times as Choice 2. The unpredictability will ensure that students consider both possibilities when they choose their tasks.

Choice 1: Find five items in the room that are between 10 cm and 25 cm long.

Choice 2: Find five items in the room that are shorter than your pencil.

CCSS: Measurement & Data: K.MD, 1.MD, 2.MD   **BIG IDEAS:** 7.1, 7.4, 7.6
Mathematical Practices: 5, 6

Some students will be comfortable with standard units of measure, whereas others will still be at the stage where comparison, rather than the use of units, is more meaningful. The two choices offered allow each group of students to attack a problem at an appropriate level.

Rulers can be provided for **_Choice 1_**, although students might first estimate without using a ruler and then check their estimates. Similarly, students might take a pencil with them as they attempt **_Choice 2_** or they might, instead, estimate and then check once they have chosen their items.

Questions that would be appropriate for both groups include:

- *What items did you find?*
- *How do you know that your items meet the required rule?*

TEACHING TIP. Questions that build on students' everyday lives help them see the value of understanding mathematical ideas.

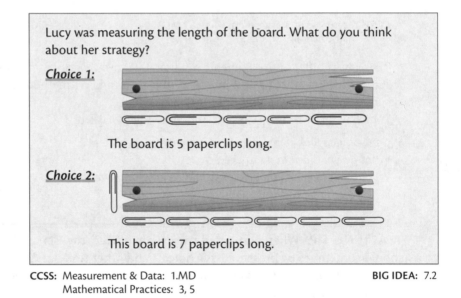

Lucy was measuring the length of the board. What do you think about her strategy?

Choice 1:

The board is 5 paperclips long.

Choice 2:

This board is 7 paperclips long.

CCSS: Measurement & Data: 1.MD
Mathematical Practices: 3, 5

BIG IDEA: 7.2

In each situation presented, errors were made in making the measurement, thus decreasing the accuracy. In **_Choice 1_**, Lucy used different unit sizes and did not measure quite end to end. In **_Choice 2_**, Lucy included more than one linear measurement in the description of the length and also extended her measurement a bit beyond the board.

Students can select the "problem" they see more easily. However, in discussing their chosen situation, students will begin to think about the other situation as well.

Questions that apply to students engaging in either task include:

- *What does it mean to find the length of the board?*
- *What parts of the board do you not care about when you measure the length?*
- *Which parts do you care about?*
- *Did Lucy measure well or did she make mistakes?*
- *How would you tell Lucy what she might do differently?*

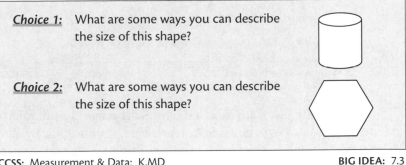

Choice 1: What are some ways you can describe the size of this shape?

Choice 2: What are some ways you can describe the size of this shape?

CCSS: Measurement & Data: K.MD   **BIG IDEA:** 7.3
Mathematical Practice: 4

Some students will be more comfortable describing the size of a 2-D shape, whereas others will be more comfortable with a 3-D shape. In the case of the hexagon, students might consider the width, the height, the lengths of **diagonals**, the space it occupies, or the length of each side. In the case of the cylinder, students might consider the height, the width, the distance around, or maybe even the thickness of the shape if it is seen as hollow.

Whichever choice students select, it would be meaningful to ask:

- *Would you consider the size large or small or neither?*
- *How could you compare the size of this shape with other shapes you know?*
- *How could you decide if your shape is bigger or smaller than another shape?*

> *Variations.* This question can be reused by specifying other pairs of shapes: one 3-D and the other 2-D, or perhaps two 2-D shapes with one more complex than the other.

Choice 1: Describe an object that is small but might weigh a lot.

Choice 2: Describe an object that is very tall, but is not very wide.

CCSS: Measurement & Data: K.MD, 1.MD   **BIG IDEA:** 7.3
Mathematical Practices: 3, 5

Both of these tasks help students to see that an individual object can be measured in more than one way. *Choice 2* involves only length measures, whereas *Choice 1* one involves both weight and another measure (e.g., length).

Providing a choice allows the student who might find it difficult to visualize a heavy or light object to focus only on length.

In either case, students can build on their everyday life experience to help them answer the question. Potential objects for *Choice 1* might be a heavy metal ball or a small rock. Objects for *Choice 2* might include a length of string or ribbon or the crack between the two parts of a folding door.

No matter which choice was selected, a student could be asked:

- *What object did you choose? How does it fit the rule for its size?*
- *Why is it probably not a good idea to tell someone "I'm thinking of something small" and assume they will know what you mean?*

Parallel Tasks for Pre-K–Grade 2

> What coin combinations can you use to show your amount?
>
> **_Choice 1:_** 12¢   **_Choice 2:_** 60¢

CCSS: Measurement & Data: 2.MD   **BIG IDEAS:** 7.4, 7.6
      Mathematical Practices: 2, 4

Frequently, in a math class, a single task is assigned to all students. For example, all students might be asked to represent 32¢ with coins. By allowing a choice that provides for different levels of complexity, more students can achieve success. The student who is really only comfortable with dimes and pennies can be as successful in his or her task as another student who can deal with a broader range of coins and more combinations.

There are only four combinations that will yield 12¢ (12 pennies, 7 pennies and 1 nickel, 2 pennies and 1 dime, or 2 pennies and 2 nickels), whereas many more combinations will produce 60¢ (e.g., 6 dimes; 5 dimes and 10 pennies; 5 dimes, 1 nickel, and 5 pennies; 2 quarters and 10 pennies; etc.).

Questions that would be appropriate to ask all students, no matter which task they chose, include:

- *Did you use any dimes?*
- *Did you need to use pennies?*
- *What coins did you use?*
- *Were there other possible combinations?*

> **_Choice 1:_** Ella has 1 quarter, 2 dimes, and 1 nickel. What's a good way to figure out how much her money is worth?
>
> **_Choice 2:_** Andrea has 3 quarters, 3 nickels, and 6 pennies. What's a good way to figure out how much her money is worth?

CCSS: Measurement & Data: 2.MD   **BIG IDEA:** 7.5
      Mathematical Practices: 2, 4, 5

Students have options about how to count the money efficiently, but in each case, they can be led to see that we usually group money to make counting/adding more efficient.

For **_Choice 1_**, students might realize the 2 dimes and a nickel are equivalent to a quarter and count 25, 50. However, they might also count 25, 35, 45, 50. For **_Choice 2_**, students might realize that a combination of a quarter and a nickel makes 30¢ and might count 30, 60, 90, 96. However, they might also count 25, 50, 75, 80, 85, 90, 96.

> **Variations.** The coin combinations can be altered to make other problems where students are likely to determine alternative efficient ways to count.

[**TO THE TEACHER:** Provide a set of sortable material. It might be **attribute blocks**; it might be buttons; it might be pictures of various colors and types of flowers.]

The teacher has set out some items that differ in various ways.

Choice 1:   Sort the items. Describe your **sorting rule**.

Choice 2:   Sort the items into two groups so that one group has two more items in it than the other group. Describe your sorting rule or rules.

CCSS: Measurement & Data: K.MD, 1.MD   **BIG IDEA:** 7.8
       Mathematical Practice: 3

These tasks differ in that students can use any sorting rule at all in the first instance, but in the second instance, they must use a rule that leads to a certain result. For example, if four green and two blue attribute blocks were provided, students could sort by color and solve either task correctly. If there were four green and four blue blocks, the students who selected *Choice 2* would need to determine a sorting rule involving something other than color.

Whichever choice was selected, in follow-up discussion students could be asked:

- *How did you sort your items?*
- *Why was your sorting rule an appropriate one for these items?*
- *Where would this object go (as another object is held up) if we used your sorting rule?*
- *How many items fit your sorting rule(s)?*

TEACHING TIP. When numbered choices are offered, the "simpler" option should sometimes be presented as Choice 1 and other times as Choice 2. The unpredictability will ensure that students consider both possibilities when they choose their tasks.

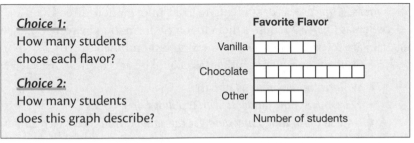

Choice 1:
How many students chose each flavor?

Choice 2:
How many students does this graph describe?

Favorite Flavor

Vanilla

Chocolate

Other

Number of students

CCSS: Measurement & Data: 2.MD   **BIG IDEA:** 7.9
       Mathematical Practice: 2

Some students will be able to read individual bits of information from the graph but will have more difficulty drawing conclusions from it, as is required to answer *Choice 2*.

Parallel Tasks for Pre-K–Grade 2

Whichever choice students selected, a teacher could ask:

- *What is this graph all about?*
- *How do you know that more students chose chocolate than vanilla?*
- *How do you know that more than nine students chose a favorite flavor?*
- *What other information does the graph tell you?*

➤ **Variations.** Any graph can be substituted that is conducive to creation of one choice that has students read information directly from the graph and another choice that has students infer information.

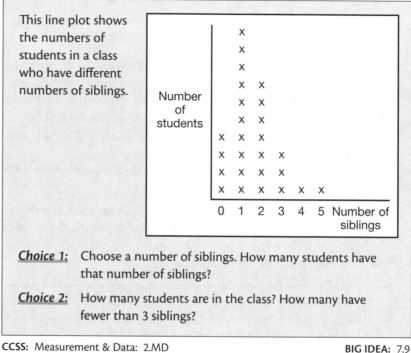

This line plot shows the numbers of students in a class who have different numbers of siblings.

Choice 1: Choose a number of siblings. How many students have that number of siblings?

Choice 2: How many students are in the class? How many have fewer than 3 siblings?

CCSS: Measurement & Data: 2.MD   **BIG IDEA:** 7.9
Mathematical Practice: 5

Both of the tasks require students to understand what a line plot is and how it works. **Choice 1** asks students to read information directly off of the line plot, whereas **Choice 2** requires them to infer information by recognizing which categories are relevant and how they need to be combined.

Students completing either task could be asked questions such as:

- *What is this line plot all about?*
- *Were there more students with 0 siblings or 1 sibling?*
- *Do you think that you would get the same data if you asked students in a different class?*
- *How many students had each number of siblings?*
- *Did any students have 6 siblings?*
- *How do you know that there were more than 10 students who answered the question?*

The final question may be somewhat more difficult than the other questions for students who have difficulty inferring information. However, it is probably simple enough (because there were 10 students in one category) that even they will be able to figure it out, and the process of working out the answer may help move them along the path toward making inferences more easily.

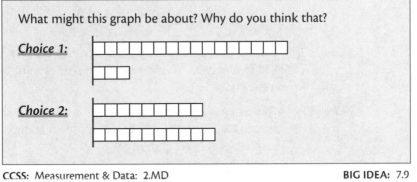

CCSS: Measurement & Data: 2.MD　　　　　　　　　　BIG IDEA: 7.9
Mathematical Practice: 3

Although students are taught how important it is to label a graph and provide titles, it can be very useful on occasion to provide an unlabeled, untitled graph for consideration. Deciding for themselves what the graph might be about, and hearing the diverse opinions of other students during class discussion, will help students understand why labels and titles are important.

For **_Choice 1_**, students need to think about a situation involving two groups that are quite different in size. For **_Choice 2_**, students need to think about a situation involving two groups that are about the same size. Students are likely to, and should, consider everyday contexts where these situations might occur.

Questions appropriate for both groups could include:

- *What title would you give the graph? Why?*
- *What labels that are missing now should be on the graph?*
- *What information does the graph tell you?*
- *Why do you think that your suggestions for the title and labels for this graph make sense?*

Parallel Tasks for Pre-K–Grade 2

PARALLEL TASKS FOR GRADES 3–5

Use this grid of dots.

. . . . . .
. . . . . .
. . . . . .
. . . . . .
. . . . . .
. . . . . .

Choice 1: Make as many shapes as you can on the grid with an area of 12 square units. The corners of the shapes must be dots on the grid.

Choice 2: Make as many rectangles as you can on the grid with an area of 12 square units. The corners of the rectangles must be dots on the grid.

CCSS: Measurement & Data: 3.MD, 4.MD   **BIG IDEAS:** 7.1, 7.4
Mathematical Practice: 5

Both of these tasks require students to consider area. Both allow for some latitude by asking students to construct as many shapes as they can, rather than a fixed number. In this way, a student who can find only one or two shapes will still feel successful.

Some students may decide not to use congruent shapes. For example, if they place a 4 by 3 rectangle in one position, they are not likely to create another 4 by 3 rectangle in another position. Encourage them to do this to help build their spatial skills.

Some students might find **_Choice 1_** easier because they have latitude in the shape they can form. They can simply connect 12 squares in some fashion to make a shape. However, it will be hard for them to be organized to produce as many shapes as possible. Other students will recognize that the rectangles required in **_Choice 2_** must be 4 by 3 rectangles and will spend their energy on ensuring that they consider all the possible placements of the rectangles.

Students could be asked:

- *How did you find your first shape?*
- *How did you use that first shape to help you get other shapes?*
- *How do you know that your shape has area 12 square units?*
- *Are there other shapes with area 12 square units? How do you know?*

> **_Choice 1_:** Some water fills a very small part of a VERY WIDE
>   container. Some water also fills most of a VERY TALL
>   container. Can you already tell which container has more
>   water in it? If you can, explain why. If you cannot, explain
>   what you would do to figure it out.
>
> **_Choice 2_:** You are trying to decide whether a VERY BIG balloon or a
>   VERY SMALL plastic ball weighs more. What would you do?

CCSS: Measurement & Data: 3.MD   **BIG IDEA:** 7.2
Mathematical Practices: 1, 3

Although one choice involves liquid volume, or capacity, and the other involves mass, both tasks require that students realize that certain aspects of the measurements of objects may be irrelevant in particular situations. In the first case, the exact values of the linear dimensions of the container are irrelevant to the problem at hand, even though they do affect the height of the water. In the second case, the relative volumes of the objects might not relate to their relative masses.

Questions that suit both tasks include:

- *What do you call the type of measurement you want to know about?*
- *Do the sizes of the containers or objects matter in deciding what you need to do? Why or why not?*
- *What strategy can you use to answer your problem?*

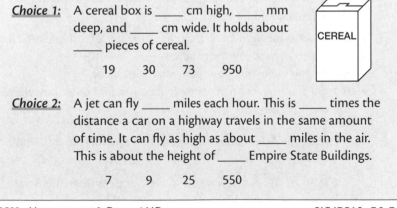

> For each puzzle, the set of all the missing numbers is listed below the
> puzzle. It is your job to figure out which number goes in which spot.
> Complete the puzzle.
>
> **_Choice 1_:** A cereal box is _____ cm high, _____ mm
>   deep, and _____ cm wide. It holds about
>   _____ pieces of cereal.
>
>   19   30   73   950
>
> **_Choice 2_:** A jet can fly _____ miles each hour. This is _____ times the
>   distance a car on a highway travels in the same amount
>   of time. It can fly as high as about _____ miles in the air.
>   This is about the height of _____ Empire State Buildings.
>
>   7   9   25   550

CCSS: Measurement & Data: 4.MD   **BIG IDEAS:** 7.3, 7.6
Mathematical Practice: 3

Both of these tasks require students to use proportional thinking. **_Choice 1_** might be easier for many students because it is a more concrete task; they can more easily imagine the size of the cereal box from their own life experience and can use the visual to help as well. On the other hand, **_Choice 1_** involves measurement

Parallel Tasks for Grades 3–5

conversions, in this case centimeters to millimeters. ***Choice 2*** might be more inter-esting to students, however, because it provides them with new information.

The solution to ***Choice 1*** is (in order) 30, 73, 19, and 950; the solution to ***Choice 2*** is 550, 9, 7, and 25.

Both groups could be asked questions such as:

- *What number were you sure of first? Why?*
- *Which number was hardest for you to get? Why?*
- *How could you be sure that your numbers made sense when you placed them in the blanks?*

➤ ***Variations.*** It is not difficult to find other objects with multiple measurements that will be familiar to students and to make up other fill-in-the-blank questions involving those measurements.

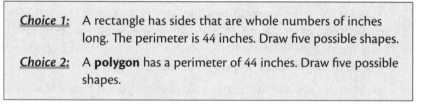

> ***Choice 1:*** A rectangle has sides that are whole numbers of inches long. The perimeter is 44 inches. Draw five possible shapes.
>
> ***Choice 2:*** A **polygon** has a perimeter of 44 inches. Draw five possible shapes.

CCSS: Measurement & Data: 3.MD, 4.MD   **BIG IDEA:** 7.3
        Mathematical Practices: 5, 6

Clearly, ***Choice 2*** provides more choice than ***Choice 1***. Some students will appre-ciate the choice, recognizing that they need only take a piece of string 44 inches long and reshape it into various polygons. Other students will feel more comfort-able working with rectangles only. Once they realize that the length and width are each represented twice in a perimeter and that therefore the length plus the width must be 22, students can simply use arithmetic skills to build their rectangles.

All students could be asked:

- *What does it mean to know that the perimeter of a shape is 44 units?*
- *How did you select your first shape?*
- *How do you know that your perimeter is 44 inches?*
- *How did you use your first shape to help you find other possible shapes?*

> ***Choice 1:*** About how much is a line of pennies that is 1 mile long worth?
>
> ***Choice 2:*** About how much is a line of pennies that is 1 yard long worth?

CCSS: Measurement & Data: 4.MD, 5.MD   **BIG IDEAS:** 7.4, 7.6
        Mathematical Practices: 2, 4

Students can line up real pennies against a ruler to get a sense of how much a line of pennies that is 1 foot long is worth. Then they can use calculations to help

Parallel Tasks for Grades 3–5

them respond to one of the given tasks. Clearly, **_Choice 2_** is easier because students need only to multiply a relatively small number by 3, rather than by 5,280.

For either choice, students could respond to questions such as these:

- *About how wide is one penny? Is that information useful? How?*
- *About how much would a line of 1,000 pennies be worth? How do you know?*
- *How could you check your results without actually lining up all the pennies?*

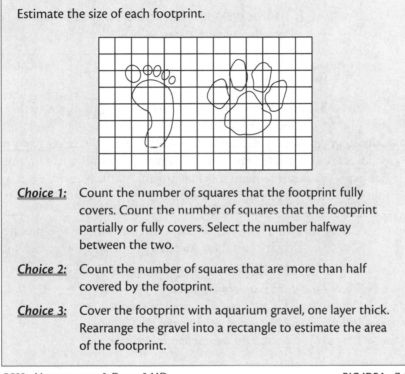

Estimate the size of each footprint.

Choice 1: Count the number of squares that the footprint fully covers. Count the number of squares that the footprint partially or fully covers. Select the number halfway between the two.

Choice 2: Count the number of squares that are more than half covered by the footprint.

Choice 3: Cover the footprint with aquarium gravel, one layer thick. Rearrange the gravel into a rectangle to estimate the area of the footprint.

CCSS: Measurement & Data: 3.MD			**BIG IDEA:** 7.4
Mathematical Practice: 5

Parallel Tasks for Grades 3–5

All three choices require students to consider what it means to estimate area. **_Choice 3_** is more labor intensive but is probably the most comfortable for students because it relates the area of something irregular to something with which they are much more comfortable, a rectangle. **_Choices 1_** and **_2_** merely require counting. **_Choice 2_** demands that students estimate whether each square is more than half covered or not; many students will find this repeated decisionmaking uncomfortable. **_Choice 1_** is somewhat more straightforward, but why a number halfway between the two values is chosen may be somewhat of a mystery.

All three groups of students could be asked:

- *Is it easy to tell which of the footprints is larger just by looking at the grid?*
- *Did your estimates help you decide which was larger?*
- *What was your estimate for each?*
- *Why did those values seem reasonable to you?*

TEACHING TIP. It is important that students get a lot of experience in measuring irregular shapes, and not just **regular shapes**. Real life is full of both.

> **_Choice 1:_** Jennifer is 5,000 days old. Describe her age in a way that is more understandable.
>
> **_Choice 2:_** Leah is 5,000,000 minutes old. Describe her age in a way that is more understandable.

CCSS: Measurement & Data: 4.MD, 5.MD   **BIG IDEAS:** 7.4, 7.6
Mathematical Practices: 2, 3

In both situations, students must use time relationships, whether it is the number of days in a year or the number of minutes in a day, week, month, or year. Although one situation is somewhat simpler than the other, both problems help students realize that the reason we use different measurement units is so that when we express a measurement, it is meaningful. In these cases, exact ages are given, but the problem is that the unit used makes the meaning of the measurement inaccessible at first glance.

Calculators should be provided as students work on this problem.

Questions that are useful for students performing either task include:

- *Why is it hard to tell how old the girl is right away?*
- *Do you think she is a baby? Explain.*
- *Do you think the girl is older or younger than you are?*
- *What unit will you use to describe her age? Why?*
- *How did you change the description to the new unit?*

> **_Choice 1:_** A structure has a volume of 80 unit cubes. It is made up of two prisms. The smaller prism is a 4 cube × 1 cube × 3 cube prism. What could the dimensions of the other prism be?
>
> **_Choice 2:_** A prism with a height of 10 cubes and a base of 6 cubes is created by putting together two other prisms that are not the same size. What could the dimensions of those prisms be? Explain.

CCSS: Measurement & Data: 5.MD   **BIG IDEAS:** 7.5, 7.7
Mathematical Practices: 1, 4, 5, 7

Both tasks involve students using the formula for the volume of a right rectangular prism.

In **_Choice 1_**, students know the volume of the smaller prism and can calculate the volume of the other. Then they use what they know about factoring to figure out possible dimensions for the second prism. Since the second volume must be 68 unit cubes, dimensions will be factors of 68, and they could have, for example, a 2 × 2 × 17 prism.

In ***Choice 2***, students can calculate the volume of the full structure and have more latitude than in ***Choice 1*** in determining the dimensions of the smaller prisms. For example, the student could decide that since the prism could be a 2 × 3 × 10 prism, it could be made up of a 2 × 1 × 10 prism *beside* a 2 × 2 × 10 prism, or perhaps a 2 × 3 × 6 prism *atop* a 2 × 3 × 4 prism.

In each case, students also use visualization skills.

Questions suitable to students engaged in either task include:

- *Is your big structure a prism or not? Does it have to be?*
- *What did you need to know about the relationship between the volume of a prism and its length, width, and height?*
- *Did you need to know the volume of your smaller prisms in order to figure out their dimensions?*

➤ **Variations.** Adjustments can easily be made in the details of either the smaller or larger prisms.

> ***Choice 1:*** A certain angle D can be broken up into three smaller angles, called angles A, B, and C. Angle B is twice as big as angle A, and angle C is twice as big as angle B. What could the measurement of angle D be? Explain.
>
> ***Choice 2:*** A certain angle D can be broken up into two smaller angles, called angles A and B. Angle B is twice as big as angle A. What could the measurement of angle D be? Explain.

CCSS: Measurement & Data: 4.MD   **BIG IDEA:** 7.5
Mathematical Practice: 4

Both of the provided tasks afford students an opportunity to think of one measure, in this case the measure of angle D, as being made up of submeasures. Both involve a bit of proportional thinking as well.

Students who attempt ***Choice 1*** might realize that angle D is actually 7 times as big as angle A. If the measurement were a whole number value, it would be a multiple of 7°. But if fractional values were allowed (and they certainly could be), angle D could be any measure at all, so long as angle A is $\frac{1}{7}$ of its size, angle B is $\frac{2}{7}$ of its size, and angle C is $\frac{4}{7}$ of its size.

Students who attempt ***Choice 2*** might realize that angle D is actually 3 times as big as angle A. If the measurement were a whole number value, it would be a multiple of 3°. But if fractional values were allowed, angle D could be any measure at all, so long as angle A is $\frac{1}{3}$ of its size and angle B is $\frac{2}{3}$ of its size.

➤ **Variations.** The relationships among the parts that make up angle D can certainly be varied.

> **_Choice 1:_** Choose the length and width for a rectangle. Now double each dimension. How does the new area compare to the old one?
>
> **_Choice 2:_** Choose the length and width for a rectangle. Now add 2 units to the length and 2 units to the width. How does the new area compare to the old one?

CCSS: Measurement & Data: 4.MD   **BIG IDEA:** 7.7
Mathematical Practices: 1, 3, 4

On the surface, these problems look fairly similar; the only difference is whether multiplication or addition was performed. Mathematically, however, the problems turn out to be quite different. In the first instance, the new area is always four times the old one; that is because, physically, the new rectangle is made up of four of the old ones.

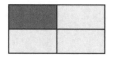

In the second instance, the relationship depends on the original length and width. Notice that in the first situation below, where the length and width are fairly large initially, the new area is barely more than the old one, but in the situation at the right, it is significantly more.

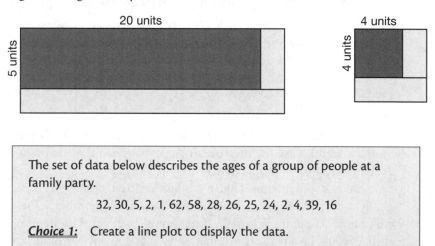

> The set of data below describes the ages of a group of people at a family party.
>
>   32, 30, 5, 2, 1, 62, 58, 28, 26, 25, 24, 2, 4, 39, 16
>
> **_Choice 1:_** Create a line plot to display the data.
>
> **_Choice 2:_** Create a bar graph to display the data.

CCSS: Measurement & Data: 3.MD   **BIG IDEA:** 7.9
Mathematical Practice: 5

Both of the suggested graph types are suitable for displaying the data given. A student is more likely to use **intervals** for a bar graph than for a line plot, and thus **_Choice 2_** may be perceived to be the more difficult task (to figure out what intervals to use). On the other hand, it may be perceived to be simpler than creating a line plot because bar graphs are more familiar to some students.

Whichever task was selected, all students could be asked:

- *How does your graph show how many people at the party are under 20? Between 30 and 40?*
- *How does your graph show the range of ages at the party?*
- *Why is your graph an appropriate way to show the data?*

Choice 1:   You collected the following data about the number of paper clips some students were able to link together in 2 minutes:

16, 18, 19, 20, 21, 21, 22, 24, 24, 25, 26, 26,
28, 29, 29, 30, 31, 32, 34, 36

Create a bar graph with four or five bars to display the data.

Choice 2:   The bar graph shown describes the number of paper clips some students were able to link together in 2 minutes. List three different sets of 20 possible data values this graph might describe.

Paper Clips in 2 Minutes

12–20   21–29   30 or more

CCSS: Measurement & Data: 3.MD
Mathematical Practice: 5   **BIG IDEA:** 7.9

Parallel Tasks for Grades 3–5

Both of these tasks require students to work with bar graphs with intervals and a scale. In one case students must create such a graph, and in the other case they must interpret a graph. By requiring four or five intervals in *Choice 1*, rather than only three, students cannot simply rely on place value concepts to define their intervals. *Choice 2* emphasizes the notion that once data have been grouped into intervals, it is no longer possible to know what the exact original data were.

Students who selected either choice could be asked:

- *Why do you think someone might use graphs that are based on intervals instead of having separate bars for each value?*
- *Are the intervals for your graph the same size? Why is that a good idea?*
- *Which interval had the greatest number of values in it?*
- *How does the graph show this?*
- *If different intervals had been used, would the graph have looked the same?*

Use the data below that describe the total number of times students in two classes had been to these places:

Disney World	25	Disneyland	11
Statue of Liberty	6	Alamo	3
Grand Canyon	5	Washington Monument	4

Choice 1: Draw a picture graph where the scale is 2, that is, one symbol represents 2 people.

Choice 2: Draw a picture graph where you choose a scale, but the scale cannot be 2.

CCSS: Measurement & Data: 3.MD   **BIG IDEA:** 7.9
       Mathematical Practice: 5

Many students default to a scale of 2 when creating a picture graph with a scale, making it easier to count and easier to use a partial symbol if required. However, often that is not the most efficient scale to use. This parallel task allows students the chance to be successful if they do choose a scale of 2, but it also encourages students who are ready to make more informed decisions about what scale would be a good choice.

Questions that would be suitable for all students include:

- *How could we use your graph to see how many more visits there were to Disney World than Disneyland in these two classes?*
- *Why might it be hard to be sure how many more of the students visited Disney World than Disneyland?*
- *Why did you choose the symbol that you did for your graph?*
- *How many symbols did you use for the Disneyland visits? Why that many?*
- *Why are there some partial symbols in your graphs?*
- *What does each partial symbol represent?*

You are going to measure 10 things at home, all between 10" and 20" long, using one of the rules below (either *Choice 1* or *Choice 2*). You will record your measurements and bring them to school. Tomorrow, in class, you will make a line plot of your measurements. Then you will create and solve two problems based on your line plot.

Choice 1: Measure to the nearest inch.

Choice 2: Measure to the nearest quarter of an inch.

CCSS: Measurement & Data: 3.MD   **BIG IDEA:** 7.9
       Mathematical Practice: 5

Although the Common Core State Standards for this grade band would expect students to measure to the nearest quarter of an inch and solve problems based on those fractions, some students whose fraction sense is poor may benefit from the opportunity to measure to the nearest inch but be part of a discussion including students who measured to the nearest quarter of an inch. Even the student selecting *Choice 1* will have experiences working with line plots and understanding what the visual display quickly communicates about the shape of the data. Allowing students to measure their own choices of items adds to the interest value as students discuss their graphs. Invariably they will talk about the items they measured and why they measured those items. The task also may provide a positive way to involve families in the student's math learning.

Questions suitable for all students involve asking them about the spread of the data, what the shape of the plot tells them about the data, what problems they posed and why those were chosen, and how they solved those problems.

TEACHING TIP. One of the most important things to consider when creating parallel tasks is that it may be necessary to use standards from earlier grades to provide success opportunities for struggling students. These students will still benefit from the discussion, which is at grade level.

SUMMING UP

MY OWN QUESTIONS AND TASKS
Lesson Goal: Grade Level: _____
Standard(s) Addressed:
Underlying Big Idea(s):
Open Question(s):
Parallel Tasks:
Choice 1:
Choice 2:
Principles to Keep in Mind:
• All open questions must allow for correct responses at a variety of levels.
• Parallel tasks need to be created with variations that allow struggling students to be successful and proficient students to be challenged.
• Questions and tasks should be constructed in such a way that will allow all students to participate together in follow-up discussions.

The nine big ideas that underpin work in measurement and data were explored in this chapter through about 75 examples of open questions and parallel tasks, as well as variations of them. The instructional examples provided were designed to support differentiated instruction for students in the Pre-K–Grade 2 and Grades 3–5 grade bands.

The examples presented in this chapter are only a sampling of the many possible questions and tasks that can be used to differentiate instruction in measurement and data. Other questions and tasks can be created by, for example, using alternate measurements or alternate graphs. A form such as the one shown here can be a convenient template for creating your own open questions and parallel tasks. Appendix B includes a full-size blank form and tips for using it to design your own teaching materials.

Geometry

DIFFERENTIATED LEARNING activities in geometry are derived from applying the Common Core Standards for Mathematical Practice to the content goals that appear in the Geometry domain for three grade bands: Prekindergarten–Grade 2, Grades 3–5, and Grades 6–8.

TOPICS

Before differentiating instruction in geometry, it is useful for a teacher to have a sense of how work in geometry develops over the grades in the curriculum.

Prekindergarten–Grade 2

Within this grade band, students begin by informally describing the shapes around them and naming 2-D and 3-D shapes. They begin to build structures and observe how large shapes or structures can be made up of smaller shapes. By combining shapes, students are building a foundation for later work on area and with fractions.

Grades 3–5

Within this grade band, students extend their use of specific properties of 2-D and 3-D shapes—such as side relationships, angle relationships, or types of **faces**—to classify them. They build, represent, and analyze both 2-D and 3-D shapes to better understand them. Students specifically consider **line symmetry** in 2-D shapes, as well as classification of shapes based on their shared attributes and the introduction of the Cartesian coordinate grid. They explore how shapes can be combined and dissected to support later work with area and fractions in two dimensions and volume in three dimensions.

Grades 6–8

Within this grade band, students focus on **composing** and **decomposing** shapes in two dimensions and three dimensions to support work in area and volume. They also solve more complex geometry problems than younger students would, including ones using the **Pythagorean theorem**, which relates lengths of sides in a **right triangle**.

Scale drawings are introduced, as are **transformations** (**rotations**, **reflections**, **translations**, and **dilatations**). Transformations also become the vehicle for exploring **congruence** and **similarity**.

Angle properties involving shapes and **parallel lines** are explored. Students perform constructions, particularly of triangles, to meet specific conditions.

Students create **nets** of 3-D shapes and explore **cross-sections** of prisms and **pyramids**.

THE BIG IDEAS FOR GEOMETRY

In order to differentiate instruction in geometry, it is important to have a sense of the bigger ideas that students need to learn. A focus on these big ideas, rather than on very tight standards, allows for better differentiation.

It is possible to structure all learning in the topics covered in this chapter around these big ideas, or essential understandings:

8.1. Different attributes of shapes and figures can be used to sort and classify these shapes in different ways.

8.2. Different tests can often be used to determine if an object is a certain kind of shape; many of these tests require measuring.

8.3. Much of what we want to find out about shapes involves measuring them.

8.4. Any shape can be represented in many ways. Each way highlights something different about the shape.

8.5. Composing and decomposing a shape can provide information about the shape.

8.6. Shapes can be transformed in a variety of ways without affecting either their size or their proportions.

8.7. Different transformations of shapes affect how the points on that shape move on the coordinate plane.

8.8. Different systems can be used to describe the location of objects.

The tasks set out and the questions asked about them while teaching geometry should be developed to reinforce the big ideas listed above. The following sections present numerous examples of application of open questions and parallel tasks in development of differentiated instruction in these big ideas across the Prekindergarten–Grade 2, Grades 3–5, and Grades 6–8 grade bands.

OPEN QUESTIONS FOR PREKINDERGARTEN–GRADE 2

OPEN QUESTIONS are broad-based questions that invite meaningful responses from students at many developmental levels.

> You are making a book about yourself. Draw a shape for the cover of your book that tells something about you. Why did you choose that shape?

CCSS: Geometry: K.G   **BIG IDEA:** 8.1
          Mathematical Practice: 5

A question such as this one provides the opportunity to explore what students know about properties of shapes. As they talk about why they chose the shape they did, students may reveal a lot about themselves (an added bonus), but it is possible to direct the conversation toward geometric attributes of shapes by reminding students to describe what aspects of the shape make the selection sensible.

The question is suitable for a broad range of students because students who need to can choose a simple shape so that they will be able to explain their choice effectively.

> Use shape stickers to make a shape picture. Try to use as many different shapes as you can. Describe your picture and the shapes in it. Tell why you used those shapes.

CCSS: Geometry: K.G, I.G   **BIG IDEA:** 8.1
          Mathematical Practice: 5

A task such as this one provides the opportunity to see what students know about shapes, and it allows the teacher to see how students combine shapes to make other shapes. As students talk about why they chose the shapes they did, the teacher can elicit their understanding of where various shapes appear in their environment and build on that.

Again, students have the option of using simple shapes if that is all they are comfortable with, or they can use very complex shapes if they wish to.

➤ *Variations.* It is possible to reuse this task by adding stipulations, for example, by indicating what shapes can be used, how many shapes can be used, what the picture must be about, and so on.

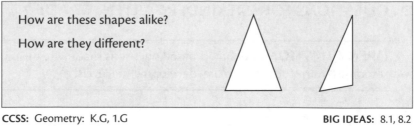

CCSS: Geometry: K.G, 1.G   **BIG IDEAS:** 8.1, 8.2
Mathematical Practices: 3, 6

This question is open in that students can choose any aspects of the shape to focus on. When they make their choice, it becomes clear what they do or do not know about triangles. Some may talk about line symmetry or lack of symmetry; some will use geometry vocabulary, for example, *They are both triangles*. Others will notice that both shapes have three sides without using the term *triangle*. Some students may even think about how the shapes can be divided or what other shape each one is a part of.

➤ *Variations.* It is possible to reuse this question with other pairs of shapes: for example, a triangle can be paired with a **quadrilateral**, a 2-D shape with a 3-D shape, or two different 3-D shapes (e.g., a cube and a long thin **rectangular prism**) can be compared.

TEACHING TIP. Some questions can be varied quite simply to be used over and over again. The question here is just one example.

Choose a type of shape. Tell as many things about it as you can.

CCSS: Geometry: K.G, 1.G   **BIG IDEAS:** 8.1, 8.2
Mathematical Practice: 6

This very open question allows students to tell whatever they know about a shape, whether it is 2-D or 3-D. Students who describe a shape reveal what they understand about the attributes or properties of the shape. Students should be encouraged to focus on geometric attributes rather than other attributes such as color or texture if the discussion begins to shift in that direction.

A student who picks a circle might say that it is round, that it is curved, or that there are no ends to it. A teacher can build on what the student has said by asking further questions. For example, a teacher might ask:

- *Can you think of a shape that does have ends?*
- *What do the ends look like?*

TEACHING TIP. One of the easiest ways to frame an open question is to mention an object, a property, a number, a measurement unit, a type of pattern, a type of graph, and so on, and ask students to tell everything they know about it.

> A certain shape makes you think of a rectangle, but it is not a rectangle. What could it be? Why?

CCSS: Geometry: K.G, 1.G, 2.G   **BIG IDEAS:** 8.1, 8.2
Mathematical Practices: 3, 6

There are many possible responses students might come up with. A few are shown here:

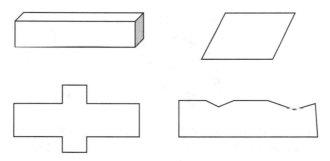

Because the question asks for a shape that makes the student think of a rectangle, there is no risk in responding—anything would be acceptable as long as students offer a reason as to why they see a connection. But the question does allow for a rich discussion of shapes in which all students can take part. Students could discuss the relationship between 2-D and 3-D shapes if a student chooses, for example, a rectangular prism. Or a student might bring out the relationship between **parallelograms** and rectangles (affording the opportunity for the teacher to explain that a rectangle is a parallelogram). Ideas of symmetry might be discussed if a student chooses a shape with the same kind of symmetry a rectangle has.

➢ *Variations.* The question can be varied by asking about other connections, for example, shapes that make the student think of a circle but are not circles.

TEACHING TIP. Differentiated instruction is supported when the environment encourages risk taking. It is only when students really feel that their teacher is open to many ideas that they start to focus on the mathematics rather than spending their time guessing what it is their teacher wants to hear.

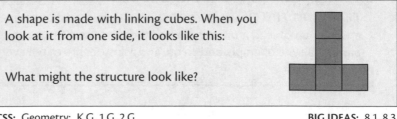

A shape is made with linking cubes. When you look at it from one side, it looks like this:

What might the structure look like?

CCSS: Geometry: K.G, 1.G, 2.G   **BIG IDEAS:** 8.1, 8.3
Mathematical Practice: 6

This task is an open one in that students are not told how many cubes to use or what the shape might look like from other perspectives. This allows all students to find an appropriate entry point. Some students will assume that the structure has only the five displayed cubes in it, whereas others will realize that there can be other cubes as well:

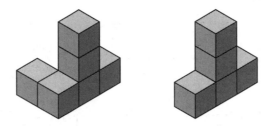

One way to vary the complexity of this question is to decide whether or not to provide linking cubes for the students to manipulate. More advanced students might be asked to visualize without actually having the cubes, although most students would benefit by having the concrete materials available.

The question can be made more complex by telling students how many linking cubes are used. If there are no conditions stated, as above, the question is much simpler than if, for example, it is stipulated that there are 11 cubes in the structure.

Build or draw a shape with 5 angles.

CCSS: Geometry: 1.G, 2.G   **BIG IDEA:** 8.2
Mathematical Practice: 5

By defining only one criterion that a shape must meet, the task is more open-ended than it might be otherwise. In this case, students must realize that if there are 5 angles, assuming a two-dimensional shape, there must be 5 sides, but the relationship between the sides is not specified. Students might be given strips of different lengths to use to create their shapes.

Some possibilities include:

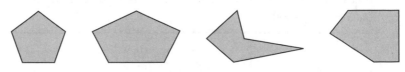

> Build or draw a shape with more than 6 faces.

CCSS: Geometry: 1.G, 2.G   **BIG IDEA:** 8.2
Mathematical Practice: 5

Again, by defining only one criterion that a shape must meet, the task is open-ended. Students might be provided with straws to build a **skeleton** or a variety of plastic shapes to serve as faces, or they might simply locate a shape in the room that meets the criterion and attempt to represent that located shape in some fashion.

Clearly the shape must be three-dimensional (since the word *faces* was used), but many shapes are possible, including hexagonal and **octagonal** prisms, hexagonal pyramids, and so forth.

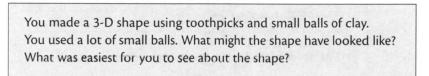

> You made a 3-D shape using toothpicks and small balls of clay.
> You used a lot of small balls. What might the shape have looked like?
> What was easiest for you to see about the shape?

CCSS: Geometry: K.G   **BIG IDEA:** 8.4
Mathematical Practice: 5

There is no restriction on the type of 3-D shape that is created, so the question is very open-ended. For example, students might create prisms, pyramids, or more irregular shapes. No matter what shape is created, students are likely to realize that what is easiest to see in this type of representation are the number of **vertices** and the number of **edges**, but seeing the number of faces is a bit more difficult.

> Draw two pictures of this box that look different.

CCSS: Geometry: K.G, 2.G   **BIG IDEA:** 8.4
Mathematical Practice: 5

Students will, of course, have varied drawing abilities, but mathematically the focus of the question is on how students might represent the shape. The fact that the shape must be drawn in two different ways will ensure that students consider what the shape will look like if it is turned or viewed from above.

Some students might draw only the front face; others might draw all six faces. Still others might try to draw a sketch that makes the shape look 3-D or might draw only the skeleton. Observing how students consider the representations of the shape will provide the teacher insight into what aspects of the shape are clear to students and what aspects need to be brought out in further instruction.

➤ *Variations.* Instead of a rectangular prism box, a triangular prism or hexagonal prism box could be used.

> Arrange tiles to form a rectangle. Make sure you use more than 30
> tiles. Is it easy to tell how many tiles make up half of your rectangle?

CCSS: Geometry: 2.G   **BIG IDEAS:** 8.4, 8.5
Mathematical Practices: 3, 5

Students are free to create a partitioned rectangle of any area greater than 30 that they wish. Some might find it easy to identify half of the rectangle because they use an even number of either rows or columns in their shapes. Others might find it more difficult.

> Use 18 linking cubes in six blocks
> of three like this one:
>
> What shapes can you make?

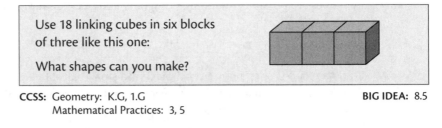

CCSS: Geometry: K.G, 1.G   **BIG IDEA:** 8.5
Mathematical Practices: 3, 5

Often geometry instruction focuses on 2-D geometry rather than 3-D geometry because that is more convenient in a classroom where there are many students but relatively few manipulatives available. However, instruction in 3-D geometry is extremely important to help students develop spatial sense. Questions such as the one above require only simple materials and are useful for geometric development.

In this case, it will be interesting for the teacher to see whether some students create only flat shapes (one layer high) or think only of traditional shapes, such as prisms. Students who are ready might use problem-solving strategies such as an organized list to come up with more options.

> **Variations.** Similar questions can be created by changing the number of small structures being combined, as well as the number of cubes or arrangement of cubes within the small structures.

> Begin with a shape of your choice. Cut it into three pieces. Give it to
> a partner to put together.

CCSS: Geometry: K.G, 1.G, 2.G   **BIG IDEA:** 8.5
Mathematical Practice: 5

Not only should students have experience composing shapes to make new ones, they should also have experience decomposing shapes into other shapes. These skills will support later work with fractions, area, and volume, and they will also help students make sense of more complex shapes they might meet.

In this situation, a student might start with, for example, a square and see that it can be cut into three pieces that are familiar shapes or three pieces that may be more unusual. Students might be surprised to learn that sometimes it is easier to put the pieces back together if the shapes are unusual (how pieces fit together is sometimes more obvious in these cases).

This task may be less intimidating to some students than other tasks would be because they have the freedom to cut wherever they wish.

TEACHING TIP. When handling discussion in a whole group setting, the teacher should call on some of the weaker students first to make sure that all of their ideas have not already been stated by the time they are called on.

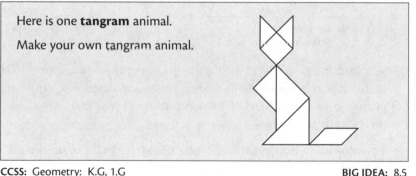

Here is one **tangram** animal.

Make your own tangram animal.

CCSS: Geometry: K.G, 1.G
Mathematical Practice: 5

BIG IDEA: 8.5

To answer this question, students would need to have been introduced to the tangram puzzle in advance. The tangram is a set of seven pieces that make a square, as shown at the right.

By providing the animal example, the question allows even the weaker student to succeed by copying the animal given. However, by opening up the task to allow students to create their own animals, all students can be appropriately challenged. Tangrams are widely used, and there are many sources for other ideas for animals, as well as for numerals, letters of the alphabet, and so on (Tompert, 1990).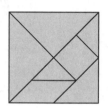

> Use any four tangram pieces to build a shape that looks like a house.
> Use geometry words to describe your house.

CCSS: Geometry: K.G, 1.G   **BIG IDEA:** 8.5
　　　　Mathematical Practice: 5

By not showing the shape the student must achieve, the question is much more open. A student can claim that his or her shape is like a house no matter what it looks like and be correct. The important part is for the student to use geometry language when he or she describes the shape. The language the student uses can be as simple or as complex as suits the individual. For example, some students will name their final shape (e.g., as a heptagon, a shape with seven sides), whereas others will simply describe the shapes that make up their house (e.g., a square and three triangles).

> Draw one picture that shows something *above* something else,
> something *next to* something else, and something *in front of*
> something else.

CCSS: Geometry: K.G   **BIG IDEA:** 8.8
　　　　Mathematical Practice: 6

Students at this level do not use grids, but they do use positional vocabulary to describe relative location. Asking for several positional situations in the same picture gives a great deal of leeway to students and certainly allows them to use their creativity.

➤ *Variations.* Other positional terms could replace the ones used in the given question.

OPEN QUESTIONS FOR GRADES 3–5

> What shapes can you build with two identical short straws and two
> identical long straws?

CCSS: Geometry: 3.G   **BIG IDEA:** 8.1
　　　　Mathematical Practice: 5

Students can be given two short and two long straws. Some of them will consider only polygons, whereas others may allow for other sorts of shapes as well.

Some students will investigate what they can do only with the given straw

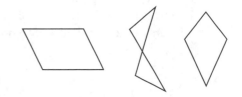

lengths. For example, given two 10-cm and two 12-cm straws, they could make a parallelogram of specific dimensions. However, other students might generalize to what happens with any two pairs of side lengths. Some students will assume that all four straws must be used, and others will consider the possibility of using only three of the straws or maybe even only two of them (to create angles). Every student in the class will be able to attack the problem at some level.

➤ *Variations.* The question can be varied by changing the number of straws or the numbers of pairs of straws that are congruent.

A 3-D shape has exactly 12 edges. What could the shape look like? How do you know?

CCSS: Geometry: 3.G, 4.G   **BIG IDEAS:** 8.1, 8.2
       Mathematical Practices: 3, 5

Visualization skills are called up on as students imagine shapes with 12 edges. The number 12 was stipulated because one of the most obvious shapes, the rectangular prism, would be a possible response for students who cannot think of other things. The interesting question becomes whether or not there are other shapes and, indeed, there are.

The base of the prism could be any quadrilateral, for example, a kite or a trapezoid, instead of a rectangle.

But the shape could also be a pyramid with a 6-sided base.

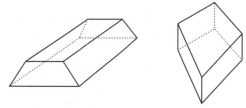

➤ *Variations.* Rather than indicating the number of edges, the number of vertices or faces could be given.

There are at least three different names you could use for a shape. What could that shape be?

CCSS: Geometry: 3.G, 5.G   **BIG IDEAS:** 8.1, 8.2
       Mathematical Practices: 3, 6

In this grade band, there is attention to the notion that some shapes are special cases of other shapes. For example, a square is also a parallelogram and a quadrilateral. (It is also a rhombus and a rectangle.) An equilateral triangle is also a triangle, a polygon, and a regular shape.

A question like this one helps students focus on the fact that most shapes have many names.

> A shape might have some pairs of **perpendicular** sides and definitely has some pairs of parallel sides. There are more pairs of parallel sides. What could the shape be?

CCSS: Geometry: 4.G, 5.G
       Mathematical Practices: 3, 6

BIG IDEAS: 8.1, 8.2

Addressing this question is likely to encourage students to think about all the shapes they know and the attributes of those shapes, making it a nice review sort of question. Students will realize that triangles cannot have parallel sides, so the shape could not be a triangle. Rectangles have two pairs of parallel sides, but four pairs of perpendicular sides, so they cannot be used. Parallelograms and trapezoids are possibilities, but not necessarily if they also have perpendicular sides. Regular octagons or regular hexagons would definitely meet the criteria.

> **Variations.** The relationship between the number of pairs of perpendicular and parallel sides could be changed.

> Prove that this shape has symmetry. You can use a **transparent mirror** or a ruler.
>
>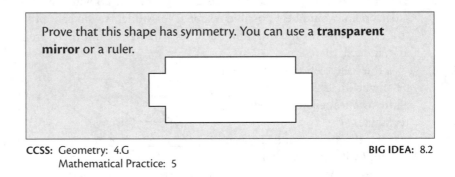

CCSS: Geometry: 4.G
       Mathematical Practice: 5

BIG IDEA: 8.2

Research has indicated that using a ruler to test for symmetry is a more developmentally demanding task than using a transparent mirror (Small, 2007). Using a ruler requires students to understand that the **perpendicular distance** to any point on one side of the **reflection line** is equal to the distance to its matching point on the other side.

By allowing students their choice of tool, the needs of students at different developmental levels are being considered.

TEACHING TIP. If students who should be working at a more challenging level consistently opt for the simpler task, the teacher should suggest to individual students or individual groups of students which tool they should use.

Exactly three sides of a shape are equal in length. What could the shape look like? How do you know?

CCS: Geometry: 3.G, 4.G, 5.G   **BIG IDEA:** 8.2
Mathematical Practices: 1, 5, 6

This question can be answered simply or in more complicated ways. Some students will immediately think of an equilateral triangle, which is a completely appropriate answer. Others will imagine shapes with more sides, only three of which are equal in length.

Some possible shapes include the ones shown below.

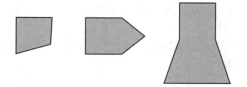

The important part of this activity is the discussion. The question is how the student created the shape and how he or she knows, for sure, that three sides, and no more than three, are identical in length. The students might be asked why particular shapes, such as parallelograms, kites, or regular hexagons, were excluded.

➤ *Variations.* The number of sides of equal length can be varied. Alternately, the number of unequal sides could be stipulated.

Fill the blanks with geometric vocabulary to make one of these statements true:

    All _____ have _____.   **OR**   All _____ are _____.

CCS: Geometry: 5.G   **BIG IDEA:** 8.2
Mathematical Practices: 3, 6

Students need to recognize that although shapes have attributes, they also have properties. When a shape has a property, there is something that is true about every example of that shape. For example, all rectangles have the property that there are 4 right angles. Properties might have to do with equal sides, equal angles, angle sizes, parallelism, and so forth.

To fill in the first set of blanks, students need to think about properties of shapes. Possible ideas are: *All rhombuses have 4 equal sides.* OR *All triangles have 3 angles.*

To fill in the second set of blanks, students need to think about hierarchies of shapes. Possible ideas are: *All rectangles are parallelograms.* OR *All triangles are polygons.*

> We use different words to describe triangles. For example, we might call them **acute**, **obtuse**, **right**, equilateral, isosceles, or **scalene**. What combinations of two words can be used to describe a triangle?

CCSS: Geometry: 5.G   **BIG IDEAS:** 8.2, 8.3
         Mathematical Practices: 5, 6

Because the question asks for which combinations are possible, and does not insist on all possible combinations, the question is appropriate for struggling learners as well as students with more developed understanding. For example, a struggling student might simply draw a triangle, notice that it is both right and isosceles and list this single combination as a possibility. Other students may want to consider all possible combinations.

In follow-up questions, students can be asked which combinations are not possible and why.

> A shape has six sides and two 90° angles. What could it look like?

CCSS: Geometry: 4.G   **BIG IDEA:** 8.3
         Mathematical Practices: 2, 5

Some possible shapes are shown below:

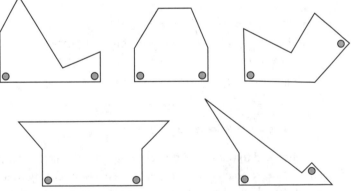

Some students will focus on starting with a line segment with two right angles at either end, then making sure there are six sides altogether, whereas others might recognize that the right angles do not need to be attached to the same side of the shape. Again, the question provides students the opportunity to offer more and less standard responses.

The teacher could follow up by asking what would happen if there were only four sides. Some students will recognize that the types of shapes then become much more limited, such as the one shown below.

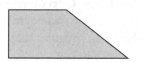

> **Variations.** The question can be varied by changing the conditions on the number of sides, the number of angles given, or the sizes of the angles given.

> You create a shape and split it up into 6 shapes of equal area. Show how this is possible.

CCSS: Geometry: 3.G   **BIG IDEA:** 8.5
     Mathematical Practices: 5, 6

Students have more options when asked to subdivide into 6 equal pieces than they would if they were asked, for example, for 7 equal pieces or even for 3 equal pieces. To get the 6 pieces requested, they might begin with a rectangle of any size, halve it and then cut the halves into thirds in a variety of ways, as shown below.

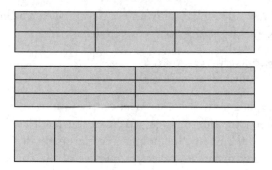

Instead of beginning with a rectangle, they might also start with a regular hexagon.

> Create a shape on a coordinate grid that has a lot of sides and one vertex at (3,1). Name the other vertices.

CCSS: Geometry: 5.G   **BIG IDEA:** 8.8
        Mathematical Practices: 4, 5

Students have a great deal of freedom in choosing the type of shape created. It might be a typical shape like a parallelogram (if a student decides that 4 is a lot of sides), but it might also be an irregular shape with any number of sides (6, 7, 9, 12, etc.), such as the one shown below.

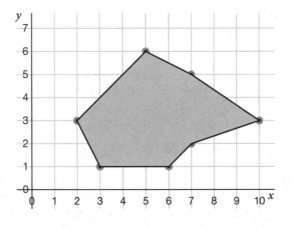

> Draw a quadrilateral that is not a rectangle on a coordinate grid. Name the vertices with ordered pairs.

CCSS: Geometry: 5.G   **BIG IDEA:** 8.8
        Mathematical Practice: 4

Students are given a great deal of freedom in deciding whether to start with the ordered pairs (four of them) to create the quadrilateral, or to start with a quadrilateral and then name the points. The placement of the quadrilateral on the grid will, in many instances, reflect properties of the shape, particularly if there are parallel, perpendicular, or equal side lengths.

> An isosceles triangle is drawn on a coordinate grid. One of the vertices is at the point (4,5). Where might the other vertices be?

CCSS: Geometry: 5.G   **BIG IDEA:** 8.8
        Mathematical Practices: 5, 6

At this level, students begin to use the first quadrant of a coordinate grid to locate objects in space. Asking students to create isosceles triangles on a grid is an interesting way to have them practice their coordinate graphing skills, while still supporting other geometric concepts.

There are many possible solutions with (4,5) as the vertex where the equal sides meet; one is shown at the left below, but the baseline could easily be extended the same amount on both sides of the vertex. However, (4,5) need not be the place where equal sides meet, as shown at the right below.

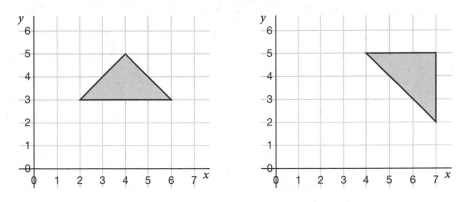

➤ *Variations.* Instead of an isosceles triangle, some other shape could easily be selected.

> A shape with some parallel sides goes through both the points (4,3) and (5,6) on a coordinate grid. What could the shape look like and what might the coordinates of its vertices be?

CCSS: Geometry: 5.G   **BIG IDEA:** 8.8
         Mathematical Practices: 5, 6

Part of the openness in this question is that the parallel lines could be horizontal or vertical (which makes the task easier) or diagonal (which makes it more challenging). For example, a student might draw a simple shape, such as the one shown at the left below. Another student might create a more complicated shape, such as the one shown at the right below.

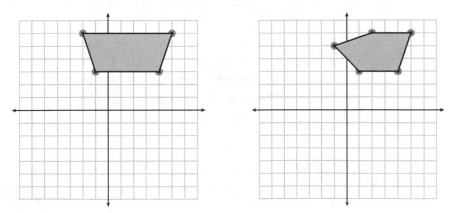

➤ *Variations.* The coordinate values can be changed, and the requirement could be for perpendicular sides rather than parallel ones.

OPEN QUESTIONS FOR GRADES 6–8

> Draw different-looking triangles where one of the angles measures 100° and one of the sides is 3" long. What else do the triangles have in common?

CCSS: Geometry: 7.G　　　　　　　　　　　　　　　**BIG IDEAS:** 8.1, 8.3
　　　　　Mathematical Practices: 1, 5

Students will, of course, draw only obtuse triangles because one of the angles must be 100°. Some will realize this right away, but others will need to actually do the drawings to see it.

Because no other angle or side length is specified, there will be great diversity in what the students draw. This experience will set them up for the realization that two pieces of information are never enough to define a triangle and that three pieces of information may or may not define a single triangle, depending on what pieces of information are offered.

> You draw a net of a 3-D shape in order to calculate its **surface area**. When you look at the net, there are only two face sizes. What shape might it be a net of?

CCSS: Geometry: 6.G　　　　　　　　　　　　　　　**BIG IDEAS:** 8.2, 8.3, 8.4
　　　　　Mathematical Practice: 3

This question is designed to focus students on what sorts of nets they might see when they want to use a net to calculate surface area.

If, for example, the shape is a rectangular prism but not a cube, a student might see only two face sizes if the base is a square. Students might also see two face sizes if the shape is any pyramid where the base is a regular shape that is not an equilateral triangle. Students might also see only two face sizes for a non-rectangular prism.

> A right triangle has two side lengths of 6" and 8". A mathematically similar triangle has one side length of 24". What could all three side lengths of each shape be?

CCSS: Geometry: 8.G　　　　　　　　　　　　　　　**BIG IDEAS:** 8.2, 8.3, 8.6
　　　　　Mathematical Practices: 3, 6

What makes this question interesting is that there is no diagram, so the student does not know which of the original side lengths matches the new 24" side length. It could be the 6" that matches the 24" and all of the original side lengths are multiplied by 4. But it could be the 8" that matches the 24" and all of the original side lengths are multiplied by 3. Or it could actually be the third side of the original right triangle, which is either 10" ($\sqrt{6^2+8^2}$) or $\sqrt{28}$" ($\sqrt{8^2-6^2}$) long, in which case the scale factor is 24 divided by either amount. Thus, the question might be viewed as either simple or complicated.

It is because the triangle is a right triangle and the Pythagorean theorem can be applied that the third side is actually known.

➤ *Variations.* Instead of using a common multiple of the two given side lengths in the smaller triangle, like 24, a number that is a multiple of only one of the provided length measures, or of neither measure, could be used.

> Create a parallelogram and a triangle so that the parallelogram area is half the triangle area.

CCSS: Geometry: 6.G   **BIG IDEA:** 8.3
Mathematical Practices: 2, 5, 7

By realizing that the area of a triangle is half of the area of a parallelogram with the same base and height, some students will realize that they can simply ensure the triangle has the same base as the parallelogram and four times the height, or the same height and four times the base, or double the base and double the height.

They might, too, use a grid rather than a formula and simply work out a possibility.

> The surface area of a rectangular prism is 300 square inches. What might the length, width, and height of the prism be?

CCSS: Geometry: 6.G   **BIG IDEA:** 8.3
Mathematical Practices: 1, 4, 7

There are many prisms that would work, so long as $lw + wh + lh = 150$, where l is the length of a base, w is the width, and h is the height of the prism. Students can select values for l and w and find the corresponding value of h, or values for l and h and find the corresponding value of w, or values for w and h and find the corresponding value of l. One example is $l = 5"$, $w = 5"$, and $h = 12.5"$.

➤ *Variations.* The value for the surface area can easily be altered.

> One of the measurements of a circle is 10 cm. What might the other measurements be?

CCSS: Geometry: 7.G   **BIG IDEA:** 8.3
Mathematical Practices: 4, 5

Students have a choice of using the 10 cm to represent the **radius**, the **diameter**, or the **circumference** of the circle. Based on that choice, other measurements will vary. It is interesting that although the 10 cm cannot be the area of the circle, the area can be determined regardless of whether the 10 cm represents the radius, the diameter, or the circumference.

> The volume of a prism is 4 cubic inches, but the length, width, and height involve fractions of inches. What could they be?

CCSS: Geometry: 6.G   **BIG IDEA:** 8.3
       Mathematical Practices: 4, 6

To solve this problem, students need to think of 4 (or a fraction equivalent to 4) as the product of three fractions. For example, if a student thinks of 4 as $\frac{120}{30}$, the dimensions could be $\frac{5}{2}'' \times \frac{4}{5}'' \times \frac{6}{3}''$ or, perhaps, $\frac{2}{5}'' \times \frac{10}{3}'' \times 3''$.

Alternately, some students can randomly choose two fractions as the length and width, divide 4 by the product of those two fractions, and then figure out the height. For example, $4 \div (\frac{3}{16}'' \times \frac{11}{4}'')$ results in a third dimension of $\frac{256}{33}''$.

> An object has a volume of 1 cubic yard. What might it be?

CCSS: Geometry: 6.G, 7.G, 8.G   **BIG IDEA:** 8.3
       Mathematical Practice: 3

By encouraging students to become aware of some familiar volume measurements, the teacher will make it easier for them to estimate the volumes of other objects. For example, if a student knows that 1 cubic yard might be the volume of a box that holds a large TV set, they will be in a much better position to estimate, for example, the volume of a room.

The question is open in that students are free to think of any object they wish. Some might think of a pile of leaves or the space inside a car or truck; or others might use yardsticks and measure a space in the classroom.

➤ *Variations.* Other values, not 1 cubic yard, can also be used.

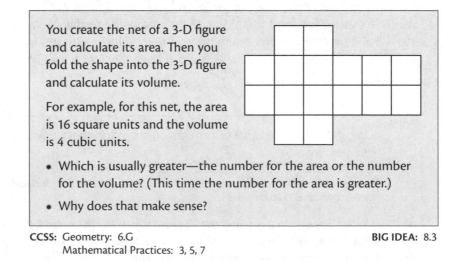

> You create the net of a 3-D figure and calculate its area. Then you fold the shape into the 3-D figure and calculate its volume.
>
> For example, for this net, the area is 16 square units and the volume is 4 cubic units.
>
> • Which is usually greater—the number for the area or the number for the volume? (This time the number for the area is greater.)
>
> • Why does that make sense?

CCSS: Geometry: 6.G   **BIG IDEA:** 8.3
       Mathematical Practices: 3, 5, 7

This open question requires students to make a conjecture. Some will try perhaps one other shape, others will recognize the value of trying many more shapes, and a few students may apply algebraic thinking to solve the problem. To think about why their answer makes sense, students need to either visualize a variety of situations or use formulas to help them understand.

Students might learn that if one of the dimensions of a prism is 1 unit, the surface area is always a greater number. Otherwise, either value can be greater. For example, for a $10 \times 10 \times 20$ prism, the volume is 2,000 cubic units, but the surface area is only 1,000 square units. For a $3 \times 4 \times 5$ prism, the volume is 60 cubic units, but the surface area is 94 square units.

> You start with a parallelogram. You increase its height by the same amount as you decrease its base length. How does the area change?

CCSS: Geometry: 6.G   **BIG IDEA:** 8.3
Mathematical Practices: 2, 5, 6

To respond to this question, some students will use a formula. For example, a student who realizes that $A = bh$ might try various number combinations for b and h to see what happens. In the table below, students can see that the area seems to go down as b and h become more different, but they are likely to try other combinations.

b	4	3	2	1
h	3	4	5	6
A	12	12	10	6

Other students will simply draw parallelograms on a grid and determine what happens to the area.

Some students will increase the height and decrease the base by a fixed amount, whether 1 unit or some other amount, whereas other students will try different amounts of increase and decrease. Some students will start with parallelograms with base and height dimensions that are close to one another, whereas others will start with a base much greater than the height or a height much greater than the base. Regardless of how many examples they try or what method they use, all students will practice the measurement of area and all will have the opportunity to learn that something different can happen depending on the value of $b - h$.

> Choose an angle size and construct that angle. What could you do to create an angle of the same size and the angle's **supplement** without using a protractor?

CCSS: Geometry: 7.G, 8.G   **BIG IDEA:** 8.3
Mathematical Practices: 5, 6

Although it may be necessary to ensure that students recognize that the supplement of an angle is the one with a measure equal to 180° less the original angle measure, students can take different directions to accomplish the task.

Some students might draw the angle and extend both rays to form both the supplement and an equal **vertical angle**.

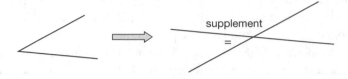

Other students will extend each ray and add a parallel line to create parallel lines cut by a **transversal**.

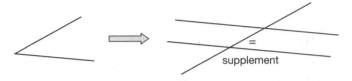

Still others might choose to create a triangle with that angle, trace it, and duplicate it to create a congruent angle and put the other two angles in the triangle together to make the supplement.

A circle is placed inside a square so that it touches the square on all four sides. About how much of the square does the circle take up?

CCSS: Geometry: 7.G   **BIG IDEA:** 8.3
　　　　 Mathematical Practices: 4, 5, 6

This question is open in that it allows a variety of approaches. Some students might use the formulas for the area of a circle and a square. They will realize they are comparing πr^2 to $(2r)^2$, so the ratio is $\frac{\pi}{4}$, or about $\frac{3}{4}$. Other students might cut out a figure like the one shown, transfer it to a grid, and count grid squares. Still other students might estimate that the pieces at the corner look like about one fourth of the total area of the square, and therefore estimate the circle to be $\frac{3}{4}$ of the total area.

Students might record the ratio as a fraction, a decimal, or a percentage. They might try many circles and squares or might assume that what is true in one case is more generally true.

TEACHING TIP. Questions that have students compare measurements are often more valuable for students than questions involving only one measurement. They get double the practice, and the notion of measurement as a comparison is reinforced.

> You are required to construct a triangle where one side is 10 cm long,
> one side is a lot longer, and one angle is obtuse. Decide first on the
> long side length and the angle measure. Then construct the triangle.

CCSS: Geometry: 7.G   **BIG IDEA:** 8.3
        Mathematical Practice: 5

Rather than instructing students as to what side and angle measures to use, there is value in letting students choose. One advantage is that students can include the angle between the two sides to make the task less difficult if they need to. That said, some students should probably be challenged to make the angle one of the non-included angles. Another advantage is that students will discover that certain combinations of side lengths may not be possible, for example, 10-30-10. Even more importantly, the diversity of triangles produced will allow for a rich class discussion.

If, for example, a student chooses a longer length of 20 cm and an angle of 120°, he or she might proceed as shown below:

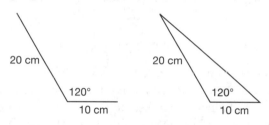

If, though, the angle is not included, a possible triangle is this one:

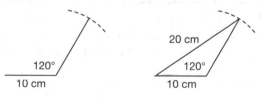

> **Variations.** Instead, the conditions might vary, for example, the student might be asked to construct an acute triangle where one side length is double another or where one angle is double another.

> You are given a shape and most of its dimensions. You need the
> Pythagorean theorem to figure out the area of the shape. What could
> the shape be and what dimensions would you already have?

CCSS: Geometry: 8.G   **BIG IDEA:** 8.3
        Mathematical Practices: 3, 6

To respond to this question, students need to visualize what sort of shape to use. Initially, most will choose a right triangle, triggered by the phrase Pythagorean theorem, but if the values of the two legs of that triangle are given (rather than the value of one leg and the **hypotenuse**), the Pythagorean theorem is actually not required to determine the area.

Students who are ready should be encouraged to think of other shapes as well. One possibility is a trapezoid like the one shown at the right, where all four side lengths are given and the height is determined using the Pythagorean theorem.

Another possibility is a regular hexagon, with the side lengths given. Again, the Pythagorean theorem is useful for determining the height of each of the six triangles making up the hexagon. This is possible because the triangles are all equilateral, so the hypotenuse lengths are also known.

Students can be challenged to think of other shapes as well.

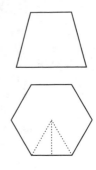

You are drawing a map of your state on a piece of notebook paper. What would be a reasonable scale for that drawing? Why?

CCSS: Geometry: 7.G   **BIG IDEA:** 8.3
Mathematical Practices: 3, 5, 6

This very open-ended question allows students to make decisions about what makes a scale reasonable, as well as doing some actual calculations to make a case for that scale. Students are likely to think of the orientation and size of the state being drawn, as well as the size of the piece of paper.

TEACHING TIP. As students are working, the teacher might ask questions such as *What makes a map on a piece of paper look "good"?*

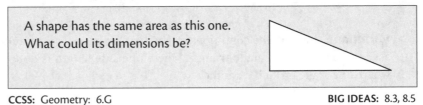

A shape has the same area as this one. What could its dimensions be?

CCSS: Geometry: 6.G   **BIG IDEAS:** 8.3, 8.5
Mathematical Practices: 4, 5, 7

Students are free to calculate the area of the triangle in a variety of ways. Some might measure the base and height and use the formula for area; others might superimpose the triangle on a grid and calculate the area that way.

Some students will realize that they do not need to measure at all. They could simply trace and cut the triangle into two pieces and rearrange the pieces to create a new shape.

Other students will think more algebraically and recognize that they do not need to measure. They could consider the formula $A = bh \div 2$ and realize they could simply double the base and halve the height to end up with the same area.

In discussing the use of the formula, the teacher could emphasize how simple it is to measure the length of the base or height as opposed to needing a grid to determine the triangle's area. This emphasis helps reinforce for students the power of measurement formulas.

➤ *Variations.* Instead of showing a right triangle, a different sort of triangle or a parallelogram might be used.

> Draw a right triangle. Build another shape on the hypotenuse so that one side of the shape is the hypotenuse of the original right triangle.
>
> Now build shapes mathematically similar to the one you created on the other sides of the original right triangle.
>
> Calculate the areas of the three shapes you created. What do you notice?

CCSS: Geometry: 8.G   **BIG IDEAS:** 8.3, 8.7
        Mathematical Practices: 3, 5

One of the interesting extensions of the Pythagorean theorem is that if similar shapes (not just squares) are built on the three sides of a right triangle, the largest area is the sum of the areas of the two smaller shapes built. By posing the question in this way, so that students can choose the shape to build, they can make the work more or less challenging, as is appropriate.

Some students might build a rectangle or a simple isosceles right triangle on each side of the original right triangle, as shown below. This makes the calculations for similarity as well as for determining area much simpler.

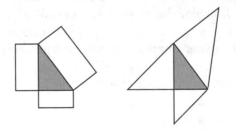

Other students might choose to put semicircles on each side of the original right triangle, and still others might choose much more challenging, unusual shapes.

> Draw several polygons on a coordinate grid to make a picture. The only conditions are that somewhere there is a side with a length of 3 coordinate units, somewhere a side with a length of 5 coordinate units, and somewhere a side with a length of 10 coordinate units.

CCSS: Geometry: 6.G, 8.G
       Mathematical Practices: 5, 6

BIG IDEA: 8.3

This question is designed to provide an opportunity for students to show their understanding of how to measure horizontal and vertical distances on a coordinate grid, although some students who are ready might choose to measure diagonal distances, either using a ruler and relating the measures to the measures on the grid or using knowledge of the Pythagorean theorem. The question is posed in such a way that students can use simple shapes if they wish or more complex ones.

Making a picture is clearly not a critical part of the mathematics learning, but it does tend to make the activity more engaging for students.

> Choose a 3-D shape. Describe all the possible cross-sections you could create.

CCSS: Geometry: 7.G
       Mathematical Practices: 3, 6

BIG IDEA: 8.4

Allowing students a choice of the shape to cut can make the problem simple or more complex. Some students might choose shapes with curved surfaces, such as spheres, cylinders, or cones, whereas others might choose cubes, square pyramids, or triangular pyramids. If the shapes are made of materials that can be cut with dental floss, students can investigate the possibilities directly, which is much clearer to some students than using drawings or simply visualizing.

> A shape has a triangle for a cross-section. What could the shape be?

CCSS: Geometry: 7.G
       Mathematical Practice: 3

BIG IDEA: 8.4

This question is suitable once students know what a cross-section is. Some students will assume that a pyramid is the logical shape to try and will find that it is not as easy as they thought to get a triangle cross-section. It is possible, for example, if a **triangle-based pyramid** is cut parallel to a base or at a vertex.

Other students will recognize that if a cut is made whenever three faces meet at a vertex, a triangle can be formed. Thus, a cube could be cut to create a triangular cross-section, as shown here.

The question is suitable for a broader group of students if modeling clay is provided for students to make shapes and dental floss to allow them to cut cross-sections.

TEACHING TIP. It might be useful to use clear plastic containers filled with water to help students see cross-sections.

Can you usually divide a shape up into one or more similar shapes?

CCSS: Geometry: 8.G   BIG IDEAS: 8.2, 8.5
      Mathematical Practices: 3, 5, 6

To approach this problem, students will need to know that shapes that are similar are enlargements of or reductions of (or identical to) a given shape. A student is likely to begin with a simple shape such as a circle and try to divide it up. They will see that it is not possible to divide up a circle into only circles that are smaller.

If they try a square, they will see that it can be divided into more squares, in more than one way:

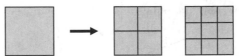

Some students will quit there, but others will investigate further, to see for which types of shapes this kind of division is possible and for which it is not. Because a simple shape like a square provides a solution, the problem should be accessible to all students in the class.

A shape is completely in the first quadrant. Where could it be after a translation?

CCSS: Geometry: 8.G   BIG IDEAS: 8.6, 8.7
      Mathematical Practices: 3, 5

Some students will try one shape and one translation and come up with an answer. For example, a student might draw a square and move it up or to the right and correctly point out that the shape has stayed in the first quadrant. Other students will realize that they could translate sufficiently to the left to end up in the second quadrant, sufficiently down to end up in the fourth quadrant, or sufficiently left and down to end up in the third quadrant. Still other students will look for all the possible combinations, showing that the shape could end up in one, two, three, or even all four quadrants.

➤ *Variations.* Similar questions can be created by changing the transformation (to rotations, reflections, or even dilatations) or by stipulating the shape with which to start.

> You draw a shape on a coordinate grid. After you rotate the shape, you draw the image. You remove evidence of the **center of rotation**, but someone else has to prove it was a rotation. What could he or she do?

CCSS: Geometry: 8.G
        Mathematical Practices: 5, 7

BIG IDEAS: 8.6, 8.7

As students are introduced to transformations, the focus should be on the effects of those transformations. Students should understand that linear, area, and angle measures are unchanged, as are relationships of parallelism and perpendicularity.

In the case of a rotation, students can observe the turn, but they need to recognize that there is a center of rotation that can be determined by locating the center of a circle on which pairs of corresponding points sit.

Using these ideas, a student can measure to see that no angle or linear measurements have changed. But the student can also locate a point equally distant from pairs of corresponding points.

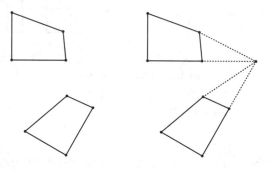

The openness in the question lies in allowing the students to choose the shape, the center of rotation, and the amount of rotation, which can simplify the task.

> **Variations.** Instead of a rotation, a reflection could be used.

> Draw a shape of your choice. Draw an exact copy of it. Prove that you can get from the first shape to the copy by using a sequence of translations, reflections, or rotations.

CCSS: Geometry: 8.G
        Mathematical Practices: 5, 6

BIG IDEAS: 8.6, 8.7

Some students may choose a very simple shape, one that is easy to reflect or rotate. Others will choose more complicated shapes. Because any of the three transformations is allowed, students will likely experience success in this task. Some students may be challenged to use only reflections to move from the first position to the second position. It turns out that this is always possible in no more than three reflections, although some students may use more.

Two examples are shown below. In the first case, an **isosceles trapezoid** can be moved by reflecting once and then translating.

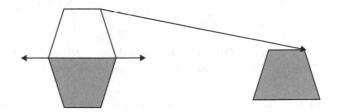

In the second case, a triangle is rotated and then translated.

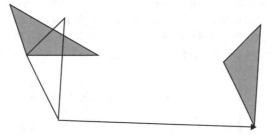

TEACHING TIP. Some students will benefit from working on a geoboard or perhaps using a transparency on which a copy of the shape can be drawn.

> Two dilatations are performed in sequence and a triangle's area becomes 36 times its original area. Show a possible set of dilatations.

CCSS: Geometry: 8.G   **BIG IDEAS:** 8.6, 8.7
Mathematical Practices: 1, 5, 6

Students might or might not choose to use a coordinate grid to perform the dilatations. They need to realize that if the area of a shape becomes 36 times as great, the linear dimensions have become 6 times as great. That means there might have been a dilatation that doubled the linear dimensions followed by one that tripled those linear dimensions. Other ratios are also possible, for example, a dilatation that took half of each linear dimension followed by one that multiplied each linear dimension by 12.

Students have the additional freedom to use any triangle they wish to show their set of dilatations.

> One vertex of a triangle is at the point (1,2). After a reflection, one
> vertex is at the point (5,8). Name all three vertices of the original and
> final triangles.

CCSS: Geometry: 8.G
Mathematical Practices: 3, 6, 7   **BIG IDEAS:** 8.6, 8.7, 8.8

Some students will only be comfortable with either horizontal or vertical
reflection lines. The way the question is posed allows them success even with this
limitation.

For example, the original triangle could have been positioned with vertices at
(1,2), (1,8), and (3,8) and the new triangle with vertices at (5,2), (5,8), and (3,8), as
shown at the left below.

Other students will assume that it was the point (1,2) that moved to (5,8). They
will look for a "diagonal" reflection line that accomplishes this task. In fact, they
could use such a line and the original vertices (1,2), (3,5), and (6,3) move to (5,8),
(3,5), and (6,3). This reflection is shown at the right below.

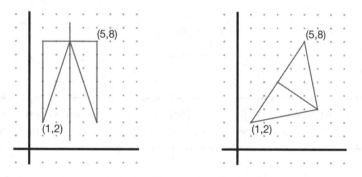

Still other students will recognize that there are even more possibilities than
that. For example, the original triangle could have been positioned with vertices at
(1,2), (0,3), and (−2,1) that move to (5,8), (6,7), and (7,10).

> Draw two similar triangles. Show the geometric transformations you
> would need to perform to prove that the triangles are similar.

CCSS: Geometry: 8.G
Mathematical Practices: 4, 5, 6   **BIG IDEA:** 8.7

An important idea for students to learn is that, just as two shapes are congruent
only if there is a set of transformations that will move one precisely to the other, the
same is true for shapes that are similar, although a dilatation must also be involved.

Some students can "play it safe" and nest one polygon into another, sharing a
vertex that turns out to be the dilatation center. Others might simply draw two
similar shapes, even in different orientations, and see that using dilatations, rotations, reflections, and translations will move one triangle to the other location.

PARALLEL TASKS FOR PREKINDERGARTEN–GRADE 2

PARALLEL TASKS are sets of two or more related tasks that explore the same big idea but are designed to suit the needs of students at different developmental levels. The tasks are similar enough in context that all students can participate fully in a single follow-up discussion.

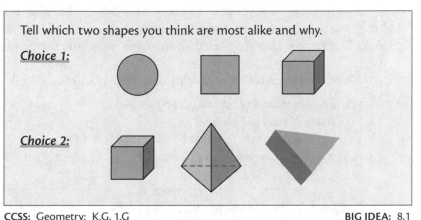

Tell which two shapes you think are most alike and why.

Choice 1:

Choice 2:

CCSS: Geometry: K.G, 1.G
Mathematical Practice: 3

BIG IDEA: 8.1

In this set of tasks, the choice is less about the developmental levels at which students are operating than it is about their comfort with a broad variety of shapes.

Some students may prefer ***Choice 2*** because they see a similarity in the shapes: all are **polyhedra** (they may not use that word, of course), or shapes with faces that are all polygons. They might focus on the fact that the second and third shapes both have triangular faces or that the first and third shapes both have rectangular faces.

On the other hand, some students might prefer ***Choice 1*** because they see two squarish shapes. Other students might see the square and circle as more similar because both are flat.

In follow-up discussion, regardless of which choice was selected, in addition to asking which shapes the student chose and why, the teacher could ask for additional observations from classmates, leading to a lively exchange to which students of all levels could contribute.

TEACHING TIP. Observing the choice a student makes when a choice is offered provides important diagnostic information to the teacher.

> **_Choice 1:_** Choose 2-D shapes to make two different creatures. Describe the two creatures you made.
>
> **_Choice 2:_** Choose 3-D shapes to make two different creatures. Describe the two creatures you made.

CCSS: Geometry: K.G, 1.G   **BIG IDEA:** 8.1
        Mathematical Practice: 5

By providing the option of working with 2-D or 3-D shapes, the problem becomes accessible to more students. Because there are no rules for what a creature looks like, there is also great freedom in which shapes are used and how they are used.

No matter which task the students selected, a teacher could ask:

- *What are the names of the shapes you used?*
- *How many of each did you use?*
- *Why did you decide those would be good shapes to use?*
- *How did you decide which shapes would be next to which other ones?*

> **_Choice 1:_** Draw a triangle that looks a little different from most triangles.
>
> **_Choice 2:_** Draw a hexagon that looks a little different from most hexagons.

CCSS: Geometry: K.G, 2.G   **BIG IDEA:** 8.1
        Mathematical Practices: 3, 5

It is important that students realize that a triangle or hexagon can be oriented in unusual positions and, as well, need not be standard in look. The task also forces students to think about what the essential features of a triangle or hexagon are.

No matter which task students select, a teacher could ask:

- *How do you know your shape is correct?*
- *What makes it different from other shapes with the same name?*

> **_Choice 1:_** A prism and pyramid have the same number of vertices. Draw what they could look like or build them.
>
> **_Choice 2:_** A prism and pyramid have the same number of edges. Draw what they could look like or build them.

CCSS: Geometry: 2.G   **BIG IDEA:** 8.3
        Mathematical Practices: 3, 5

Providing students with small balls of clay or marshmallows to serve as the corners, and straws or toothpicks to serve as the edges, enables them to model the shapes by building skeletons.

Parallel Tasks for Pre-K–Grade 2

Students who select _**Choice 1**_ will likely build the prism first. Then, once they realize that there is only one corner at the top of the pyramid, they will know exactly how many corners the base must have. For example, if a student builds a **triangle-based prism** (which has six corners), he or she will realize that a **pentagon** must form the base for the pyramid. If students start with the pyramid first, the problem may be more challenging unless they happen to realize that it is essential to start with an even number of corners on the pyramid.

Students who select _**Choice 2**_ may begin with the prism or the pyramid. Because it turns out that the number of edges on a prism is always triple the number on the base and the number of edges on a pyramid is always double the number on the base, students will discover that the problem can only be solved if the number of edges used is a multiple of both 3 and 2.

Both of these problems are challenging but manageable. The second is clearly more complex than the first.

No matter which task was selected, students could be asked:

- _How would you name your prism and pyramid?_
- _How did you count the corners or edges?_
- _How could you have predicted that your shape would have that many corners or edges?_

Pick one of these shapes. Represent it as many ways as you can. Use pictures, words, numbers, and objects.

**Choices:**   cube   square   cylinder   hexagon

CCSS: Geometry: K.G, 1.G, 2.G   **BIG IDEA:** 8.4
Mathematical Practice: 5

In posing these parallel tasks, the teacher should have a variety of materials available, including molding clay, drawing materials, paper that can be cut or rolled, pattern blocks that can be used or traced, 3-D models, straws and connectors, and **polydrons** (plastic shapes that can be connected to form 3-D shapes). These can be supplemented with other found items in the classroom; for example, a student might use a book with a front cover in the shape of a square as a model for the square. The choice in this task allows students more comfortable with 2-D to work in that realm, whereas other students can choose 3-D shapes.

A student who selects the cube might simply find models of cubes in the classroom (e.g., dice), might draw a picture of a cube, or might draw all of the individual faces of a cube. A student who chooses the hexagon may not initially realize that there are many types of hexagons—not all of them the standard regular hexagon—but may come to recognize that fact as he or she searches for six-sided objects. A student who chooses the cylinder might roll a piece of paper to make a cylinder, thereby setting up a future connection that will be useful in exploring the surface area of cylinders.

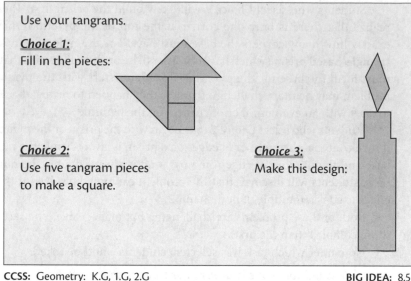

Use your tangrams.

Choice 1:
Fill in the pieces:

Choice 2:
Use five tangram pieces
to make a square.

Choice 3:
Make this design:

CCSS: Geometry: K.G, 1.G, 2.G   **BIG IDEA:** 8.5
Mathematical Practice: 5

Choice 1 is most suitable for students whose geometric concepts are less highly developed. If the diagram is made life size, students can simply experiment and try different pieces in different spots. In discussion at the conclusion of the work, those students are able to talk about how they fit the shapes in, what clues they used, and so on.

Choice 3 is a mathematically more sophisticated activity than *Choice 1* because no internal lines are presented. Its complexity, too, can be varied based on whether the diagram is life size. A life-size diagram will always be easier to work with because students will be able to fit pieces into the outline. The diagram at the right shows the tangram candle with outlines of the component pieces.

Choice 2 is the most open. An even more open alternative would be to simply ask students to use any number of tangram pieces to make a square, offering some very simple solutions. By requiring exactly five pieces, the task poses more challenge to students than it would in its simplest form. The diagram above shows a square made up of five tangram pieces, with outlines.

The follow-up discussion could invite students, no matter what task they chose, to describe how they went about solving their problem.

> *Variations. Choice 1* can be adjusted to be slightly more difficult by using a life-size diagram with no internal lines shown.

Parallel Tasks for Pre-K–Grade 2

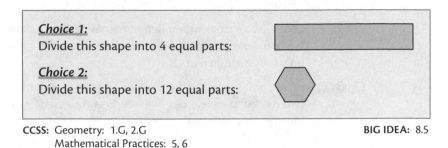

Choice 1:
Divide this shape into 4 equal parts:

Choice 2:
Divide this shape into 12 equal parts:

CCSS: Geometry: 1.G, 2.G   **BIG IDEA:** 8.5
Mathematical Practices: 5, 6

Many students will find it easier to divide the rectangle than the hexagon; they can think about folding the rectangle, but might find folding the hexagon more complex. On the other hand, the hexagon task is not overly difficult because the student might realize that he or she could first create the six triangles that make up the hexagon and then divide each one. Recognizing this possibility is more likely if students have had prior experience working with pattern blocks.

Both tasks focus on the idea that a shape can be divided into many smaller shapes. The follow-up for either task could involve asking students to talk about alternate ways to divide the shape.

➤ *Variations.* The tasks can be adapted by using other shapes and other numbers of parts.

PARALLEL TASKS FOR GRADES 3–5

Choice 1: Describe and apply two different strategies you could use to draw a pair of parallel lines. How do you know, for sure, that the lines are parallel?

Choice 2: Describe and apply two different strategies you could use to draw a pair of perpendicular lines. How do you know, for sure, that the lines are perpendicular?

CCSS: Geometry: 4.G   **BIG IDEA:** 8.2
Mathematical Practices: 4, 5

The choice provided might appeal to students' personal notions of whether parallelism or perpendicularity is easier to create. A variety of methods can be employed, including tracing available materials, using a **straightedge** and **compass**, using **dynamic geometry software**, or using a transparent mirror.

The important part of the discussion is how the students know that their lines are either parallel or perpendicular. The teacher can ask for evidence beyond "*They look right.*" The evidence will likely involve measuring distances or measuring angles. For example, two lines can be tested for parallelism by drawing a line that looks perpendicular to both and checking to see if it is. Two lines can be tested for perpendicularity by using one line as a mirror line to see if the image of one side of the other line matches the other side of the other line.

> <u>**Choice 1:**</u>   Create two shapes with different names that each have two pairs of equal sides. Tell how else they are alike and how they are different.
>
> <u>**Choice 2:**</u>   Create two shapes with different names that each have two pairs of parallel sides. Tell how else they are alike and how they are different.

CCSS: Geometry: 4.G, 5.G   **BIG IDEAS:** 8.1, 8.2
        Mathematical Practice: 5

There are several options for each task. For the first task, a student might create a rectangle and a parallelogram, a "house"-shaped pentagon and a kite, or perhaps an irregular pentagon and an irregular seven-sided shape.

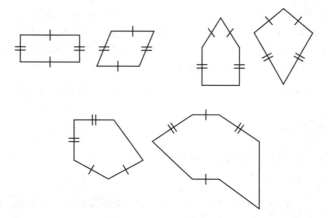

For the second task, a student might use a parallelogram and a rectangle, a seven-sided shape and a square, or perhaps a rhombus and a long parallelogram.

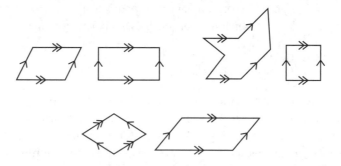

Questions suitable for both tasks include:

- *What names would you give to your two shapes?*
- *How are your shapes similar beyond having the pairs of parallel (or equal) sides?*
- *What about your shapes makes them seem the most different?*

These two dots are the
corners of a given shape:

●

                                        ●

Choice 1:   Make a square with these corners.

Choice 2:   Make a parallelogram with these corners.

Choice 3:   Make an isosceles triangle with these corners.

CCSS: Geometry: 3.G, 4.G   **BIG IDEAS:** 8.2, 8.3
          Mathematical Practice: 5

Some students will favor *Choice 3* because they only need to place one more point. The fact that the triangle must be isosceles makes the task appropriately challenging, however.

Some students who choose *Choice 1* will choose to make the dots the endpoints of a diagonal so that the sides can be horizontal and vertical. Others will realize that the square can be on a slant and will use the dots as the endpoints of one side.

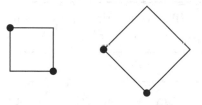

Still other students will see the slant inherent in the segment joining the two points and use horizontal lines to make the parallelogram specified in *Choice 2*.

No matter which task was selected, the student could be asked:

- *Describe your shape.*
- *What name would you give your shape? Is it a . . . ? [e.g., a polygon]*
- *How do you know it is the shape you say it is?*

> _**Choice 1:**_ A rectangle on a coordinate grid has two vertices at (1,5)
> and (3,5). Where could the other vertices be?
>
> _**Choice 2:**_ A rhombus on a coordinate grid has two vertices at (1,5)
> and (3,5). Where could the other vertices be?

CCSS: Geometry: 5.G   **BIG IDEAS:** 8.2, 8.3, 8.8
Mathematical Practice: 5

Using a coordinate grid is new to students in this grade band. It makes sense to combine the use of the grid with exploring other geometric concepts as well. In this case, students are combining the study of properties of shapes with the use of the grid. They need to address what makes a rectangle a rectangle or a rhombus a rhombus in order to figure out where the other vertices are.

Questions that suit both tasks include:

- _Why might the shapes be either big or small?_
- _Could the shapes be oriented in different directions?_
- _Are you sure where the lines of symmetry of the shapes are?_

TEACHING TIP. Using a grid to prove geometric properties in earlier grades will help prepare students for later work they meet in secondary schools.

> _**Choice 1:**_ Draw a rectangle. Divide it into four sections of $\frac{1}{4}$. Do it
> two more times, but divide it a different way each time.
>
> _**Choice 2:**_ Draw a rectangle. Divide it into four sections of $\frac{1}{4}$ so that
> the sections are not all exactly the same.

CCSS: Geometry: 3.G   **BIG IDEA:** 8.5
Mathematical Practices: 5, 6

Each choice presents an interesting challenge for students. In the first case, the student might fold horizontally twice, vertically twice, or horizontally and vertically to create the four sections. Some students might incidentally also solve the problem posed in _**Choice**_ 2, but they might not. In the second case, the student cannot do the obvious and must think about other options. A few possibilities are shown here:

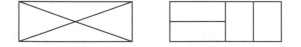 

Questions a teacher might ask could include:

- _Why is each section called $\frac{1}{4}$?_
- _How can you be sure they are the same size (area)?_
- _What process did you use? Why was that a good idea?_

> ***Choice 1:*** Show how to build a rhombus by putting together three
> quadrilaterals. Name the shapes you used.
>
> ***Choice 2:*** Show how to build a rectangle by putting together three
> quadrilaterals. Name the shapes you used.

CCSS: Geometry: 3.G, 4.G   **BIG IDEA:** 8.5
Mathematical Practice: 5

The focus of these activities is on decomposing shapes, but at the same time students think about the different types of quadrilaterals there are. In naming the quadrilaterals they use, a teacher can draw out the fact that many quadrilaterals have several names and can explore, with students, the hierarchy of quadrilateral types. The second task might be viewed as more accessible because students are much more familiar with the rectangle shape.

In responding to ***Choice 1***, a student might use three parallelograms to create the rhombus (or might call it two rhombuses and a parallelogram) or perhaps an irregular quadrilateral (the chevron) and two trapezoids.

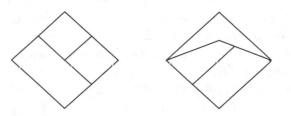

In responding to ***Choice 2***, a student might use three smaller rectangles or perhaps a square and two other rectangles.

Questions suitable for both tasks include:

- *How would you name your three quadrilaterals?*
- *What strategy did you use to break up your original shape?*
- *Are there other possible ways to break up your shape into quadrilaterals? How do you know?*

TEACHING TIP. Although it might benefit some students to have concrete shapes to work with, this level of specificity does limit students in terms of the sizes and variations of size that they can use.

PARALLEL TASKS FOR GRADES 6–8

Each prism has whole number unit side lengths.

Choice 1:   A prism has a surface area of 126 square units and a volume of 90 cubic units. What could the dimensions be?

Choice 2:   A prism has a volume of 48 cubic units and a width of 4 units. What could the dimensions be?

CCSS: Geometry: 6.G, 7.G   **BIG IDEAS:** 8.2, 8.3
       Mathematical Practice: 4

Choice 2 will be easier for many students than *Choice 1* because one of the three dimensions is provided. Students choosing *Choice 2* will realize that the product of the length and height must be 12 square units and can simply list some possibilities. The challenge of *Choice 1* requires students to simultaneously apply formulas for both surface area and volume.

Regardless of which choice the students selected, it would be meaningful to ask:

- *What is the relationship between the dimensions and the volume?*
- *What does the volume tell you about the shape?*
- *Why would there have been even more possibilities if only the volume had been given?*
- *Can you be sure whether or not there are other possible dimensions?*

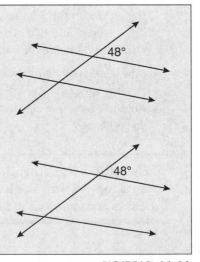

Choice 1:
Two of the lines shown are parallel. One angle measure is provided. Determine all of the other angle measures in the picture without using a protractor.

48°

Choice 2:
Two of the lines shown are almost, but not quite, parallel. One angle measure is provided. Estimate all of the other angle measures in the picture without using a protractor.

48°

CCSS: Geometry: 7.G, 8.G   **BIG IDEAS:** 8.2, 8.3
       Mathematical Practice: 7

Many students are unaware that one of the reasons we explore the angle measure as a transversal cuts a pair of parallel lines is as a test for parallelism. If students understand the relationship of angles when the transversal cuts a pair of parallel lines, they should be successful with *Choice 1*. What makes *Choice 2* interesting, though, is that they have to extend their thinking so that their angle choices are reasonable.

Questions that might be asked of both groups of students include:

- *Which angle did you figure out first? Why that one?*
- *Which angles in the picture must be equal? Which can't be equal?*
- *Can you be sure of all of your angle measures? Why or why not?*

Choice 1: One circle has a greater area than another. Does its circumference also have to be greater?

Choice 2: One rectangle has a greater area than another. Does its perimeter also have to be greater?

CCSS: Geometry: 8.G   **BIG IDEAS:** 8.2, 8.3, 8.7
Mathematical Practices: 3, 5

Some students who are new to the formulas involving circumference and area of a circle are likely to be attracted to *Choice 2*, using rectangles. What is interesting, however, is that the answer to *Choice 1* is clearer than the answer to *Choice 2*. If one circle has a greater area than another, it has a greater radius and therefore a greater circumference. On the other hand, a rectangle with a greater area can sometimes have a greater perimeter and sometimes not. For example, a 6 by 10 rectangle has more area but less perimeter than a 1 by 30 rectangle, but a 4 by 5 rectangle has more area and more perimeter than a 1 by 2 rectangle.

Students completing either task could be asked:

- *How many combinations did you try?*
- *Can you be sure of the result with that many trials?*
- *How could a picture help someone understand your thinking?*
- *How might you have predicted what you found out?*

Choice 1: You want to prove that two triangles are similar. You know the angle and length measurements of the first one. What is the fewest number of measurements you need in the second one? How do you know?

Choice 2: You want to prove that two rectangles are similar. You know the angle and length measurements of the first one. What is the fewest number of measurements you need in the second one? How do you know?

CCSS: Geometry: 8.G   **BIG IDEAS:** 8.2, 8.3, 8.7
Mathematical Practices: 1, 5

The intention of this question is to bring out the idea that only two angle measurements are required to decide if two triangles are similar, but that length measurements, and not angle measurements, are required to decide if two rectangles are

Parallel Tasks for Grades 6–8

similar. Of course, length measurements can be used with the triangle, as well, but the point is that they are not essential.

Regardless of the task selected, students could be asked:

- *How many measures are there associated with your second shape?*
- *Are there some measures you know automatically if you know other measures? Which ones?*
- *Why do you have to take measurements in both shapes?*
- *Which measurements did you decide you absolutely needed? Why those?*

TEACHING TIP. When students are given some choice in which task to perform, they are more likely to persevere with the task than when they are not given a choice.

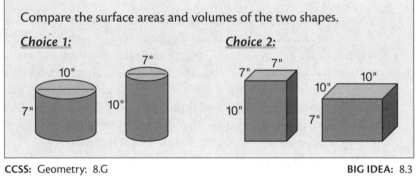

Compare the surface areas and volumes of the two shapes.

Choice 1: 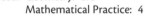   *Choice 2:*

CCSS: Geometry: 8.G
Mathematical Practice: 4

BIG IDEA: 8.3

In both cases, students need to know how to calculate surface area and volume, but in one case they must be able to deal with cylinders and in the other case only with rectangular prisms. Many students will assume that the volumes are equal in **Choice 1** because they see 7 and 10 on both shapes, but this is not the case. They may be less sure of how the surface areas compare.

In fact, for **Choice 1**, the volumes are 175π cubic inches for the short cylinder compared to 122.5π cubic inches for the tall one, and the surface areas are 120π square inches compared to 94.5π square inches. For **Choice 2**, the volumes are 490 cubic inches and 700 cubic inches, respectively, and the surface areas are 378 square inches and 480 square inches, respectively.

Whichever choice was selected, a teacher could ask:

- *Is it possible for the volumes to be the same but the surface areas to be different?*
- *Is it possible for the surface areas to be the same but the volumes to be different?*
- *Are either the same in your pair of shapes?*
- *How did you calculate the volumes?*
- *How did you calculate the surface areas?*

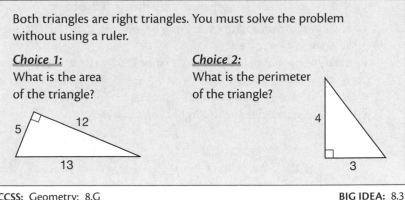

Both triangles are right triangles. You must solve the problem without using a ruler.

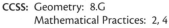

Choice 1:
What is the area of the triangle?

Choice 2:
What is the perimeter of the triangle?

CCSS: Geometry: 8.G   BIG IDEA: 8.3
       Mathematical Practices: 2, 4

Choice 1 is set up with all of the required information provided, as long as students recognize that orientation is irrelevant in describing the base and height of a triangle. Here the base could be viewed as the side with length 5 and the height as length 12; the area, therefore, is 30 square units. *Choice 2* requires application of the Pythagorean theorem.

Whichever task was completed, students could be asked:

- *You calculated one measure of the shape. How did you do the calculation?*
- *Why does that method make sense?*
- *What if you had been asked to calculate a different measure? Would it have been equally easy?*

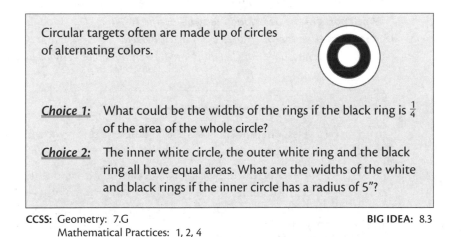

Circular targets often are made up of circles of alternating colors.

Choice 1: What could be the widths of the rings if the black ring is $\frac{1}{4}$ of the area of the whole circle?

Choice 2: The inner white circle, the outer white ring and the black ring all have equal areas. What are the widths of the white and black rings if the inner circle has a radius of 5"?

CCSS: Geometry: 7.G   BIG IDEA: 8.3
       Mathematical Practices: 1, 2, 4

Both of these choices require that students use the formula for the area of a circle. There are many more possible answers for *Choice 1* than for *Choice 2* partially, because no specific sizes are listed and partially because there are no requirements for how the two white sections relate. Because there are more potential answers, *Choice 1* may be more accessible to students. However, some students may prefer *Choice 2* because it is specific.

Whichever choice was selected, a teacher could ask:

- *How could the area of the black ring be thought of as the difference between the areas of two circles?*
- *Could the area of the outer white ring be thought of as the difference between the areas of two circles?*
- *Could the area of the inside white circle be half of the area of the whole circle?*
- *What did you decide the widths of the rings were?*

Choice 1: Create a triangle where one side is 4" long, one side is 6" long, and there is an angle of 60°.

Choice 2: Create a triangle where one angle is 40°, another angle is 60°, and the side between those angles is 4" long.

CCSS: Geometry: 7.G   **BIG IDEA:** 8.3
Mathematical Practice: 5

The two choices differ in that, in the first case, there is not a unique triangle determined because the 60° angle can be between the 4" side and the 6" side or not. In the second case, only one triangle is possible.

In both instances, students receive practice in drawing triangles to meet given conditions, but in one case there is more to explore than in the other. The discussion can bring out how certain conditions force a particular triangle and certain conditions do not.

Questions a teacher might ask could include:

- *Which parts of the description did you use first? Why?*
- *What were the other measures of your triangle?*
- *Do you think another triangle would have been possible? Why or why not?*

A prism has a volume of about 200 cm³.

Choice 1: Describe the dimensions of a different prism with about the same volume.

Choice 2: Describe the dimensions of a cylinder with about the same volume.

CCSS: Geometry: 8.G   **BIG IDEA:** 8.3
Mathematical Practice: 4

Both choices require that students have an understanding of the formula for the volume of a prism, but **Choice 2** also requires that students have either a formal or informal understanding of the formula for the volume of a cylinder. A student who chooses **Choice 1** could look for groups of three numbers that multiply to produce 200, for example, 10 by 20 by 1, or 4 by 10 by 5. A student who chooses

Choice 2 might recognize that a prism with a base area of 20 cm² and a height of 10 cm would have the desired volume. They might then choose a cylinder with a slightly larger base and similar height, figuring that the cylinder needs to be a bit wider to make up for the fact that a circle takes up less space than a square of the same width.

Both groups of students could be asked:

- *Can you be sure of the dimensions of the original prism? Why not?*
- *Is there a maximum height it could be? A minimum height? Why not?*
- *Is it helpful to know volume formulas to help you solve the problem? How?*

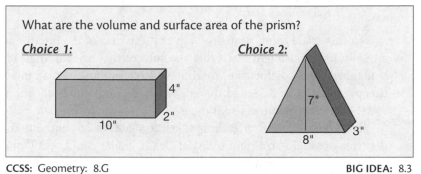

CCSS: Geometry: 8.G   **BIG IDEA:** 8.3
            Mathematical Practice: 4

To complete *Choice 2*, students not only need to know what volume and surface area mean and how to calculate them, but they also need to be able to use the Pythagorean theorem to determine the side length of the side rectangles. For *Choice 1*, they need to know how to calculate surface area and volume of rectangular shapes only. The volume and surface area for *Choice 1* are 80 cubic inches and 136 square inches; the volume and surface area for *Choice 2* are 84 cubic inches and 128.4 square inches.

Whichever choice was selected, a teacher could ask:

- *Which face has the greatest area?*
- *How do you know?*
- *Why is it hard to tell which has the greatest volume by just looking at the pictures?*
- *Does a shape that is taller always have a greater volume? A greater surface area?*
- *How did you calculate the volume? The surface area?*

> **_Choice 1:_** One of the angles in a triangle is less than 40°. Another **interior angle** is double that size. What could be the measure of all three angles in the triangle? What sort of triangle could it be?
>
> **_Choice 2:_** One of the angles in a triangle is less than 40°. An **exterior angle** is double that size. What could be the measure of all three angles in the triangle? What sort of triangle could it be?

CCSS: Geometry: 8.G   **BIG IDEA:** 8.3
Mathematical Practice: 4

In this grade band, students learn that the sum of the interior angles of a triangle is 180° and might learn that the sum of the exterior angles is 360°. Some students are less comfortable working with exterior angles than interior angles, so this parallel task is designed to support those students. Both questions have an equal degree of openness about them.

For **_Choice 1_**, the student might select angles of 30° and 60° and realize that the triangle would be a right triangle. Or the angles could be 39° and 78°; the last angle would be 63°, and the angle would be an acute triangle. The triangle cannot be obtuse, and the students would need to reason about why.

For **_Choice 2_**, the student might select an interior angle of 30°, an exterior angle of 60° (creating an interior angle of 120°), and a final angle of 30°; an isosceles obtuse triangle is created. In fact, neither an acute triangle nor a right triangle can be created.

Students can be asked about their strategies and about both the interior and exterior angle measures no matter which task they pursue.

> **_Choice 1:_** The volume of a cylinder is about 20 cubic inches. What could its radius and height be?
>
> **_Choice 2:_** The volume of a cone is about 20 cubic inches. What could its radius and height be?

CCSS: Geometry: 8.G   **BIG IDEA:** 8.3
Mathematical Practice: 4

Many students are more comfortable with cylinders than with cones; this parallel task affords them the opportunity to choose the shape that makes them more comfortable. But the tasks are so similar that assigning them at the same time should lead to no difficulties.

Questions a teacher might ask all students include:

- *Can your shape be very short?*
- *Could your shape be very tall? If it is tall, can it also be wide?*
- *Which did you decide on first—the radius or the height? Could you have done it the other way?*

> **_Choice 1:_** A net made up of 2 triangles and 3 rectangles has a total
> area between 100 in^2 and 200 in^2. Draw the net and
> indicate the dimensions. Prove that the total area is in the
> correct range.
>
> **_Choice 2:_** A net made up of 4 triangles and 1 rectangle has a total
> area between 100 in^2 and 200 in^2. Draw the net and
> indicate the dimensions. Prove that the total area is in the
> correct range.

CCSS: Geometry: 6.G   **BIG IDEAS:** 8.3, 8.4
Mathematical Practice: 5

Using nets is an effective approach to determining surface area so that the student ensures that all faces are accounted for. For **_Choice 1_**, students must realize that the shape could be a triangular prism and must recognize that, in that case, the triangles would be the same size and the rectangles could be different sizes. The shape for **_Choice 2_** might be a rectangular pyramid with pairs of triangle faces equal. Students must also apply what they know about calculating the areas of triangles.

Questions could focus on how students knew what the 3-D shape looked like, which (if any) faces they believed had to have the same size, and which measurements they decided on first and why.

> What fraction of the area of the rectangle is black?
>
> **_Choice 1:_**   **_Choice 2:_**
>
>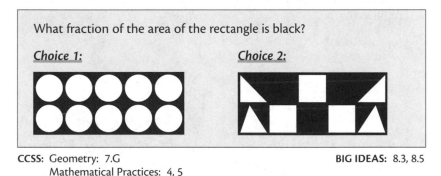

CCSS: Geometry: 7.G   **BIG IDEAS:** 8.3, 8.5
Mathematical Practices: 4, 5

Students need to know how to calculate the area of a circle to complete **_Choice 1_** (the fraction of black is actually about $\frac{1}{4}$) but only the areas of triangles and squares to complete **_Choice 2_** (the fraction of black is actually $\frac{1}{2}$). In either case, students can apply their knowledge of area formulas; all that differs is which formulas they must focus on.

Whichever choice was selected, a teacher could ask:

- *Is the black area more or less than half the total area? Why do you think that?*
- *What formulas did you use to help you?*
- *How much of the rectangle do you think is black?*

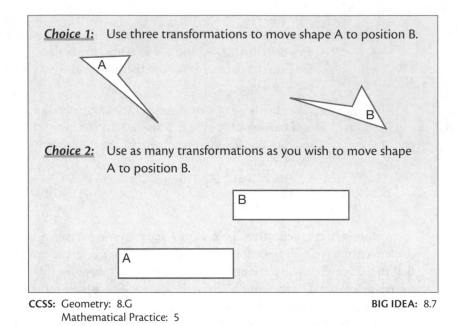

CCSS: Geometry: 8.G   **BIG IDEA:** 8.7
Mathematical Practice: 5

Both tasks require students to use transformations. **_Choice 1_** stipulates the number of transformations to be used, which adds to the complexity of the problem. Three reflections can be used to accomplish the first task; the first moves one point on the original shape to its image, and the next two together result in the required turn:

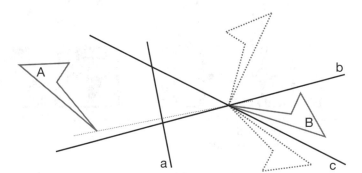

Whichever task was performed, in follow-up discussion students could be asked how they recognized which transformations to try and how they tested their predictions. Further, they could be asked to describe, using appropriate mathematical language, the transformations they used.

SUMMING UP

```
          MY OWN QUESTIONS AND TASKS

Lesson Goal:                  Grade Level: _____

Standard(s) Addressed:


Underlying Big Idea(s):


Open Question(s):



Parallel Tasks:
Choice 1:


Choice 2:



Principles to Keep in Mind:
  • All open questions must allow for correct responses at a
    variety of levels.
  • Parallel tasks need to be created with variations that allow
    struggling students to be successful and proficient
    students to be challenged.
  • Questions and tasks should be constructed in such a way
    that will allow all students to participate together in
    follow-up discussions.
```

The eight big ideas that underpin work in geometry were explored in this chapter through more than 80 examples of open questions and parallel tasks, as well as variations of them. The instructional examples provided were designed to support differentiated instruction for students in the Pre-K–Grade 2, Grades 3–5, and Grades 6–8 grade bands.

The examples presented in this chapter are only a sampling of the many possible questions and tasks that can be used to differentiate instruction in geometry. Other questions and tasks can be created by, for example, using alternate shapes or measurements. A form such as the one shown here can be a convenient template for creating your own open questions and parallel tasks. Appendix B includes a full-size blank form and tips for using it to design your own teaching materials.

Statistics & Probability

DIFFERENTIATED LEARNING activities focused on statistics and probability are derived from applying the Common Core Standards for Mathematical Practice to the content goals in the Statistics & Probability domain for Grades 6–8.

TOPICS

Before differentiating instruction in work with statistics and probability, it is useful for a teacher to have a sense of how data and probability topics develop over the grades.

Prekindergarten–Grade 2

Within this grade band, students work with data by sorting objects and answering questions about those sorts.

Grades 3–5

Students within this grade band explore data mostly through measurements they collect. They interpret and answer questions about the collected data.

Grades 6–8

Within this grade band, there is significantly increased attention to statistics and probability. Students think about what statistical questions are. They consider issues of sampling and compare **univariate data** distributions, often using measures of **center**, such as mean and **median**, and of **spread**, such as **inter-quartile range**, **mean absolute variation**, and **standard deviation**. They begin to consider the relationship between two variables, using measures of association and linear models. New data displays at this level include **dot plots**, **histograms**, **box plots**, **scatter plots**, and **two-way tables**.

Probability is formally introduced in this grade band. Students initially consider **experimental probability**, and move to **theoretical probability** involving single and then **compound events**. They also design and use **simulations** to estimate probabilities.

THE BIG IDEAS FOR STATISTICS & PROBABILITY

In order to differentiate instruction in statistics and probability, it is important to have a sense of the bigger ideas that students need to learn. A focus on these big ideas, rather than on very tight standards, allows for better differentiation.

It is possible to structure all learning in the topics covered in this chapter around these big ideas, or essential understandings:

9.1. Sometimes a large set of data can be usefully described by using a **summary statistic**, which is a single meaningful number that describes the entire set of data. The number might describe either the values of pieces of data or how the data are distributed or spread out.

9.2. Sometimes a set of data or a relationship between data can be estimated using an algebraic or graphic representation.

9.3. Sometimes data about an entire population can be estimated by using a **representative sample**.

9.4. An experimental probability is based on past events, and a theoretical probability is based on analyzing what could happen. The value of an experimental probability approaches that of a theoretical one when enough random samples are used.

9.5. In probability situations, one can never be sure what will happen next. This is different from most other mathematical situations.

9.6. Sometimes a probability can be estimated by using an appropriate model and by conducting an experiment.

The tasks set out and the questions asked about them while teaching statistics and probability should be developed to reinforce the big ideas listed above. The following sections present numerous examples of application of open questions and parallel tasks in development of differentiated instruction in these big ideas across the Grades 6–8 grade band.

OPEN QUESTIONS FOR GRADES 6–8

OPEN QUESTIONS are broad-based questions that invite meaningful responses from students at many developmental levels.

> The mean of a set of numbers is 8. What might the numbers be?

CCSS: Statistics & Probability: 6.SP   **BIG IDEA:** 9.1
Mathematical Practices: 1, 4, 5

This is a very open question. Students are free to select how many numbers to use in their set as well as the size of the numbers. Struggling students might pick a very simple set such as {7, 8, 9} or even the single number 8. But other students might select much larger data sets with more diverse values.

By providing sets of linking cubes, the teacher can make the task accessible to all students. Students might begin with any number of groups of 8 and simply move cubes around to discover new sets with the required mean. In the example shown below, the original set of {8, 8, 8, 8} cubes has been rearranged to create the new set of {7, 10, 10, 5} (mean 8).

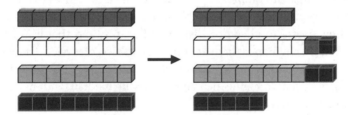

In class discussion of the different data sets students generate, the teacher can point out how, in each case, the sum of the data values is a multiple of 8. In fact, the multiple of 8 that the data values total equals the number of values in the set. For example, the sum for {7, 8, 9} is 24, which is 3 × 8 (i.e., the number of values in the set × the mean value for the set).

> **Variations.** Rather than indicating a particular value for the mean, a range of values could be allowed, for example, a mean between 10 and 20.

TEACHING TIP. Use of manipulatives is good for all students and can provide valuable scaffolding for differentiated instruction.

> Create a situation where you think the median is a better description
> of a data set than the mean. Tell why it is better.

CCSS: Statistics & Probability: 6.SP   **BIG IDEA:** 9.1
          Mathematical Practices: 2, 3, 4

Students have a great deal of freedom in developing a situation that fits the
rule here. Most likely all students will think of a situation where there are extreme
values at one end of the distribution that make the mean unusually high or low.
But the nature of the situation and how many extreme values there are is not
specified and students might think of some interesting real-life examples.

> Create a set of data in which the mean is greater than the median.

CCSS: Statistics & Probability: 6.SP   **BIG IDEA:** 9.1
          Mathematical Practices: 1, 4

Once students have become familiar with concepts of both the mean and the
median, they can handle a task such as this one. The question is open in that it
allows students to operate at their own comfort and knowledge levels to create
either simple or complex sets of data. Yet it still requires that all students deal with
the concepts that a median is the middle piece of data, that a mean is a description
of how a data total can be distributed evenly, and that a mean is more likely to
exceed the median if there are only one or two pieces of data with high values.

To very simply perform the task, a student might begin with a set of data such
as {4, 5, 6}, for which the mean is equal to the median, and then substitute a number
greater than 6 for the 6.

➤ *Variations.* Students could be asked to create data where the mean is a specified
amount more than the median, or perhaps only slightly more than the median.

> Jane and Missy are in two different classes. Each had a history test.
> Describe possible data showing the marks in each class if Jane's mark
> of 83% might actually be viewed as better than Missy's mark of 90%.
> Explain what about your data makes this a reasonable claim.

CCSS: Statistics & Probability: 6.SP   **BIG IDEA:** 9.1
          Mathematical Practices: 1, 3

In examining a set of data, students need to consider many factors, including
the center of the data and the spread of the data. A student can make choices about
what it means to be "better," but will still have to describe either the spread or the
center of the data. The openness is not only in the interpretation of what "better"
means but also in creating the data set.

As students share their thinking, they could be asked about **outliers** in the data
and about means, medians, inter-quartile ranges, and/or standard deviations.

> A box plot looks like this: ————————————————————[ ]——————
> What could the data have been?

CCSS: Statistics & Probability: 6.SP
Mathematical Practices: 2, 6

BIG IDEA: 9.1

Often an open question is the perfect way to determine whether students have a conceptual understanding of an idea. In this case, the facts that the whiskers are of different lengths and that the box is narrow provide a way to see if students really understand that a box plot is a picture of how data are distributed.

In the particular case shown, students should realize that the middle 50% of the data lie in a quite narrow range relative to the range of the whole data set. They should also realize that the lowest 25% of the data are much more spread out than the upper 25% of the data.

> **Variations.** Different-shaped box plots could be used instead. The middle box could be narrower or wider than the one shown, or the whiskers could be equal in length, or the upper whisker could be considerably longer than the lower one. An alternate box plot that is either more symmetric or even more extremely asymmetric could be used.

TEACHING TIP. Often an open question with no numbers provided is an excellent tool for assessing understanding.

> You gather data about how long it takes various people of your age to walk 1 mile. How spread out might the data be? What might cause some of the data to be very unusual? How should that affect your sample?

CCSS: Statistics & Probability: 6.SP
Mathematical Practices: 2, 3, 6

BIG IDEAS: 9.1, 9.3

The notion of outliers is an important one for students to consider as they examine data that have already been gathered, but might be considered even before the data are gathered, in terms of whether outliers should be avoided or should be deliberately included. Although this question is framed around a particular type of data, there is still some latitude in terms of what factors students might think of and how that might influence their sample.

TEACHING TIP. Questions can and probably should vary in the degree of openness. It is usually a good idea to have a mix of open questions— some that are very, very open and others that are somewhat less open— to accommodate the differences among students' comfort with these types of questions.

Create a scatter plot that relates two variables that shows a VERY negative association. What might these variables be? Why does it make sense that the association might be negative?

CCSS: Statistics & Probability: 8.SP
Mathematical Practices: 2, 4, 5

BIG IDEA: 9.2

It is likely that students will think of real-life situations where one variable is likely to decrease when the other one increases. For example, the number of hours spent playing video games might be negatively associated with academic performance. Students then create data that make sense for the given variables and justify their choice.

There are likely students who will simply draw a scatter plot with a negative association and not associate the variables with anything real. This is an entry point, so it should not be disallowed, but the teacher should encourage students, if they can, to consider what those variables might represent.

> *Variations.* Instead of a negative association, no association might be suggested.

A scatter plot has a couple of outliers, but the two variables that are graphed seem to have an almost linear relationship. Draw two possible scatter plots.

CCSS: Statistics & Probability: 8.SP
Mathematical Practices: 4, 5, 6

BIG IDEA: 9.2

This question requires students to think about what outliers are and what a **linear association** means. At the same time, by stipulating that the association be "almost linear," the question makes students think more deeply about linearity; they have to consider both when things are linear and when they are not. The request that two scatter plots be drawn prompts students to think harder about options they might have.

Create a two-way table of frequencies that might suggest that students who play a lot of sports do better in school. Then create a different table where the relationship does not seem as strong.

CCSS: Statistics & Probability: 8.SP
Mathematical Practices: 2, 4, 5

BIG IDEA: 9.2

Allowing students to decide how many data to present and what the data actually look like provides opportunities for students at different levels of mathematical comfort to access the question. Some students might use a relatively small set of data, whereas others might use a much more fulsome set of data. There may be students who need clarification of what a two-way table of frequencies is, but this question would not likely be posed until that type of information had been presented to students.

It would be interesting to see how students view a strong or not as strong relationship, and this question will help make that apparent.

> The line $y = 4x - 10$ describes a scatter plot fairly well, but not exactly. Create a table of values to show the possible points on the plot.

CCSS: Statistics & Probability: 8.SP   **BIG IDEA:** 9.2
Mathematical Practices: 4, 5

Most students are likely to graph the line $y = 4x - 10$ and then move a few points up or down off the line, and this is an appropriate response. The teacher might encourage some students to start with the table of values directly, rather than the graph. Those students might realize that, although they might normally quadruple the x value and subtract 10, they won't always do that.

It would be interesting to discuss how varied the responses to this actually are. Do most students stay very close to the line or do some veer off more?

> **Variations.** A line with a negative slope might be suggested instead.

> Think of a situation you are interested in where someone might use information from a survey to make a decision that affects many more people than were surveyed. For example, a commercial gym might ask just some clients about what hours they would prefer and assume other clients would agree. Once you have decided on your situation, discuss how many people and what people you would survey to have a representative sample.

CCSS: Statistics & Probability: 7.SP   **BIG IDEA:** 9.3
Mathematical Practices: 3, 6

The notion of sampling is something that is increasingly part of every citizen's need to understand his or her surroundings. With the Internet, society is bombarded with data, often from samples, all the time. Critical questions are always how the data were collected and whether the sample is representative.

To better engage students, they can be encouraged to investigate a topic of interest to them rather than a topic predetermined by the teacher. The important concept dealt with in this question, regardless of the topic investigated, is what makes a sample representative. Students will need to consider either sample size or sample constitution or both. A class discussion will, no doubt, be interesting because of the variety in what will be discussed.

Open Questions for Grades 6–8

> In testing voters' opinions, one pollster got results that showed 35%
> of the people in a certain area supported a particular political party.
> Another pollster got results that showed that 37% supported that
> party. A third got results showing that 52% supported that party.
> How could that happen? What can you conclude?

CCSS: Statistics & Probability: 7.SP   **BIG IDEA:** 9.3
       Mathematical Practice: 3

Most students will initially suggest that the last pollster must be wrong since the other two are close. This can lead to an interesting discussion about whether this conclusion is actually warranted.

Encourage students to think of lots of reasons why results might vary, for example, how many people were polled, the "randomness" of the make-up of the samples, or other factors such as time of polling and so forth.

Hopefully, students realize that one can't conclude much without a lot more information.

> Imagine that a new student is about to join the class. Decide which
> of these statements is likely, which is certain, which is unlikely, and
> which is impossible.
>
> - The student is a boy.
> - The student is the same age as many other students in the class.
> - The student is 20 years old.
> - The student has a head.
> - The student lives in the local area.
> - The student likes school.
> - The student has very few close friends.
>
> Add your own statements that other students can judge as likely,
> unlikely, certain, or impossible.

CCSS: Statistics & Probability: 7.SP   **BIG IDEA:** 9.4
       Mathematical Practices: 3, 4

When first learning about probability, students are much more successful using qualitative terms such as *likely, unlikely, certain,* and *impossible* rather than numerical ratios or fractions. The task posed in this question is accessible to many students because it is grounded in their everyday experience. It also allows for differentiation because the students are asked to make up their own statements. The statements they make up could have to do with age, gender, habits, preferences, family relationships, and so on.

➤ *Variations.* The task can be varied by offering statements about a sports figure or musical artist, but care must be taken to not make the question inaccessible to some students by requiring too much specialized knowledge.

> Describe an event that fits each word:
>
> Impossible   Certain   Unlikely

CCSS: Statistics & Probability: 7.SP   **BIG IDEA:** 9.4
       Mathematical Practice: 3

With such a general question, students have a lot of opportunity to be success-ful. They can draw on personal, everyday experience or school knowledge. For example, a student might say that it is certain that an even number does not end in 3, or that it is unlikely that a number less than 100 describes the number of people in a town.

> A **tree diagram** showing the outcome of an experiment has 12 branches when it is complete. To what problem might the tree diagram represent the solution?

CCSS: Statistics & Probability: 7.SP   **BIG IDEA:** 9.4
       Mathematical Practices: 2, 4, 5

Naturally, this question is only appropriate once students have become familiar with tree diagrams. With that knowledge base in place, the question is open because the complexity of the tree diagram that is considered is up to each individual student.

Some students will draw a tree diagram representing a simple experiment, for example, spinning a spinner with 12 sections labeled **A–L** and letting each branch of the tree diagram represent one of those 12 outcomes. Other students will think of 12 as the product of 2 and 6 and use a tree diagram to represent an experiment involving two aspects, one of which has 2 possibilities and one of which has 6 pos-sibilities, for example, flipping a coin and rolling a die.

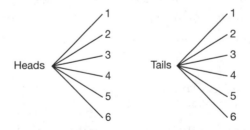

Still other students will think of a combination of two events with 3 and 4 pos-sible outcomes, or a combination of three events with 2, 2, and 3 possible outcomes, for example, flipping two coins and spinning a spinner with three equal sections.

The discussion about the diverse tree diagrams and problems created by the students will be a rich one and will benefit a broad range of individuals.

➤ *Variations.* The number of branches can be varied, although it is helpful to ensure that it is a number with several factors.

Open Questions for Grades 6–8

> Imagine that you are selecting colored cubes out of a bag. What
> colors of cubes and how many of each color might be in the bag if
> the probability of selecting a green one is close to $\frac{1}{3}$?

CCSS: Statistics & Probability: 7.SP   BIG IDEA: 9.4
       Mathematical Practices: 2, 3

Students have a great deal of freedom to determine the various colors of cubes
in the bag as well as the numbers of cubes. The response could range from something as simple as a bag with three cubes, where only one is green, to a situation in
which there are 100 cubes, 30 are green, and the rest are distributed evenly among
four other colors.

In addition, students are allowed to decide what "close to" means in this context. Some might argue that a 1 in 4 probability is close enough, whereas other
students might demand a fraction that is closer to 1 in 3. The mathematical concept
of estimation is an important one for all students. Class discussion of this task will
help students understand that estimation is not a science, but an art.

TEACHING TIP. There can never be too much discussion about estimation.
It is important for students to become comfortable with deciding when to
estimate and what ranges of estimates are appropriate in different situations.

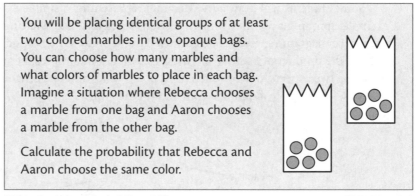

> You will be placing identical groups of at least
> two colored marbles in two opaque bags.
> You can choose how many marbles and
> what colors of marbles to place in each bag.
> Imagine a situation where Rebecca chooses
> a marble from one bag and Aaron chooses
> a marble from the other bag.
>
> Calculate the probability that Rebecca and
> Aaron choose the same color.

CCSS: Statistics & Probability: 7.SP   BIG IDEA: 9.4
       Mathematical Practices: 2, 4

Students select the number of marbles in each bag as well as their colors. The
response might be as simple as placing two blue marbles in each bag; then the
probability of both students choosing the same color would be 100%. Many students, however, will select marbles of different colors. For example, if 2 green and
3 blue marbles were placed in each bag, the probability of selecting two marbles of
the same color would be $\frac{13}{25}$ (i.e., the probability of selecting two greens, $\frac{2}{5} \times \frac{2}{5} = \frac{4}{25}$,
plus the probability of selecting two blues, $\frac{3}{5} \times \frac{3}{5} = \frac{9}{25}$).

TEACHING TIP. When a sufficiently open question is asked, the discussion will be a rich one because there will be so many directions in which students might go and so many possible results to discuss. Inevitably, students will learn more about the concept at hand than if a more closed question is asked.

> You have just rolled one 6-sided die ten times. Your results were: 5, 5, 5, 1, 5, 1, 2, 4, 5, 5.
>
> How likely do you think it is that you will get a 5 on the next roll? Explain.

CCSS: Statistics & Probability: 7.SP   **BIG IDEA:** 9.5
      Mathematical Practice: 3

One of the most uncomfortable aspects of probability for students is that what you expect does not always happen. Some students will look at the rolls above and not believe it could happen; how could you have so many fives? Others will assume that with all of those fives, there is no way of getting another five for a while. Still others, ideally most, will realize that you really can't be sure what will happen next. The probability for a five is still $\frac{1}{6}$, like it always was.

> An experiment involves using this spinner and die. Create a situation for which the combined probability is $\frac{1}{2}$ or more.
>
>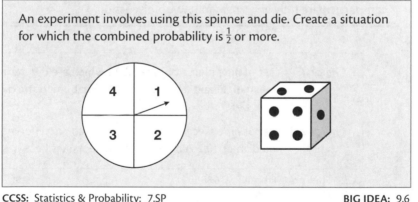

CCSS: Statistics & Probability: 7.SP   **BIG IDEA:** 9.6
      Mathematical Practices: 2, 5

This question is open in that students have a fair bit of latitude in what the probability could be. They also have freedom in deciding what the events might be. For example, a student might decide that the spinner could land on 2 or 3 (which has a probability of $\frac{1}{2}$) along with any value on the die. The combined probability would be $\frac{1}{2}$ (i.e., $\frac{1}{2} \times 1$), and the problem would be solved. Another student might decide that the combined probability of landing on a number less than 4 on the spinner and a number less than 6 on a die (i.e., $\frac{3}{4} \times \frac{5}{6} = \frac{15}{24}$) is a possible solution. A very simple solution would be allowing for any roll and any spin, for a combined probability of 1, which does satisfy the problem requirements.

Open Questions for Grades 6–8

> A certain pizza place advertises a special deal if you let them choose the toppings. They use a random 3 ingredients out of the following: pepperoni, mushroom, sausage, green pepper, ham, pineapple, ground beef, and onion. How could you set up an experiment to see the probability that you will get pepperoni on each of the next three pizzas you order?

CCSS: Statistics & Probability: 7.SP   **BIG IDEA:** 9.6
        Mathematical Practices: 4, 5

Simulations are a useful tool for students to solve real-life problems involving probability. In this situation, students need to think about how to best model "mathematically" the real-life situation. Students need to realize that some device with 8 equal sections is needed and then need to think about how to model buying three pizzas (most likely selecting 3 out of 8 cubes 3 times or spinning 3 sections on an 8-section spinner, etc.).

PARALLEL TASKS FOR GRADES 6–8

PARALLEL TASKS are sets of two or more related tasks that explore the same big idea but are designed to suit the needs of students at different developmental levels. The tasks are similar enough in context that all students can participate fully in a single follow-up discussion.

> _**Choice 1:**_   A set of nine pieces of data, all of which are different, has a mean of 30 and a median of 10. What could the data values be?
>
> _**Choice 2:**_   A set of nine pieces of data, all of which are different, has a median of 10. What could the data values be?

CCSS: Statistics & Probability: 6.SP   **BIG IDEA:** 9.1
        Mathematical Practices: 4, 5

For _**Choice 2**_, students need only to realize that the median is the middle number of the data set. They could simply write down the number 10 and list four different numbers less than 10 and four different numbers greater than 10. For _**Choice 1**_, the student must negotiate two pieces of information. Not only does the median have to be 10, but the mean has to be 30. The student might realize that the sum of the data values must be 270 (i.e., 9 values × 30 for the mean) or might simply guess and test to find values that work.

In a discussion about the tasks, all students will demonstrate their understanding of the concept of median, but even the students who elect **_Choice 2_** will be reminded about how means are calculated. Students who complete either task could be asked:

- *How do you know your median is 10?*
- *What is the least value you had? Did it have to be less than 10? Why?*
- *What was the greatest value you had? Did it have to be greater than 10?*
- *How did you determine your set of data values?*

Choice 1: Create a set of five pieces of data with a mean of 6.
No more than one of the values can be 6.

Choice 2: Create a set of five pieces of data including the values
4, 2, and 2, and where the mean is 20.

CCSS: Statistics & Probability: 6.SP   **BIG IDEA:** 9.1
Mathematical Practices: 5, 6

Both tasks require that students understand the concept of mean. **_Choice 1_** is less restrictive. Students can use any values they wish to create the data set. A student with a good understanding of the concept of mean might begin with {6, 6, 6, 6, 6} and increase two of the values and decrease two others in the same way. This approach might yield a solution of {2, 4, 6, 8, 10} (increasing and decreasing values by 2 and 4). **_Choice 2_** requires the student to recognize the need for greater numbers in the remainder of the data set. He or she might recognize that the five values must total 100 (i.e., 5 values × 20 for the mean) and thus that the two remaining values must total 92 (i.e., 100 − 4 − 2 − 2). Another student might employ an alternate strategy.

No matter which choice or approach was selected, students could be asked:

- *What is a mean?*
- *How do you know that your mean is correct?*
- *How did you figure out the other numbers?*

TEACHING TIP. This question was set up to require working backward. Rather than giving data and asking for the summary statistic, the statistic is provided and the data are requested. This approach results in a more open question.

Choice 1: Create two sets of 12 data values each so that the mean deviation for the first set of data is 3 more than for the second set of data.

Choice 2: Create two sets of 12 data values each so that the mean deviation for the first set of data is 3 times as much as for the second set of data.

CCSS: Statistics & Probability: 6.SP
Mathematical Practices: 4, 6

BIG IDEA: 9.1

Some students who select *Choice 2* will realize that tripling the set of data provides a simple way to satisfy the requirement; others will attempt to meet the condition in more unusual ways. Students who select *Choice 1* might originally decide to add 3 to each data value (which will not change the mean deviation) or to one or two data values (which will also not solve the problem). Those students will have to do more thinking about how to solve the problem.

Students working on either task could be asked:

- *How do you calculate the mean deviation?*
- *Could you have changed just one data value? If so, how?*
- *Is the problem solved simply by changing all the data values? If so, how?*
- *What does a greater mean deviation actually mean? Does your second data set reflect this?*

Choice 1: Create a set of data that can be appropriately described by using a histogram. Create that histogram.

Choice 2: Create a set of data that can be appropriately described by using a box plot. Describe that plot.

CCSS: Statistics & Probability: 6.SP
Mathematical Practice: 5

BIG IDEA: 9.2

Both choices require students to consider a type of graph and when it is best used. In the case of the histogram (*Choice 1*), the student needs to recognize that the data should describe a continuous measure where it makes sense to count frequencies of data in different intervals. For example, the data could be heights of students in different age groups. In the case of the box plot (*Choice 2*), the student needs to recognize that the data must be organized into quartiles and the middle 50% of the data must be separated from the top and bottom 25%.

Regardless of which choice a student selected, a teacher could ask:

- *How did you organize your data?*
- *Why did it make sense to use that type of graph for these data?*
- *Looking at the graph, what picture of the data does it give the viewer?*

> Two teenage girls sell jewelry. Their monthly income for each of eight months is shown below.
>
> | Month | Sept | Oct | Nov | Dec | Jan | Feb | Mar | Apr |
> |---|---|---|---|---|---|---|---|---|
> | Income, in $ | 45 | 60 | 85 | 110 | 130 | 145 | 170 | 190 |
>
> **Choice 1:** Create a scatter plot and describe how the income seems to change as the number of months grows.
>
> **Choice 2:** Create an equation that you think closely represents the relationship between the income and the number of months the girls have been selling jewelry.

CCSS: Statistics & Probability: 8.SP   **BIG IDEA:** 9.2
         Mathematical Practices: 4, 5

Choice 1 might be more attractive to students who choose to graph and esti- mate a rough **line of best fit**, while *Choice 2* might be more attractive to students who are comfortable with algebraic/technological methods for determining lines of best fit. No matter which choice students make, they are exploring the descrip- tion of a set of data mathematically.

Questions suitable to students in both groups include:

- *Do you think that the income values will continue to rise? Can you be sure?*
- *Is the rise from month to month relatively constant? How can you tell from your graph or your equation?*
- *Why might it be useful to the girls to have a description of this relationship?*

> The line $y = 3x + 2$ is a good, but not perfect, model to describe how two sets of data are related. What could the data values in each set be?
>
> **Choice 1:** Determine the data values by using a graph.
>
> **Choice 2:** Determine the data values without using a graph.

CCSS: Statistics & Probability: 8.SP   **BIG IDEA:** 9.2
         Mathematical Practices: 4, 5

Often, **bivariate data** are provided and students are asked to determine a line of best fit. By providing the equation of a line of good fit, but not best fit, and asking for the data, the teacher gains greater insight into student understanding of what a line of fit does.

The look-fors here include whether the data represented by the y-values increase at about three times the rate as the data represented by the x-values, how

Parallel Tasks for Grades 6–8

the *y*-intercept of 2 is made reasonable, and how the student ensures that the line is not a perfect fit. *Choice 1* allows students to use a graph; typically students plot the graph and then move a few points off of the graph. *Choice 2* forces students to think more algebraically about what the relationship actually means.

Questions that suit both tasks include:

- *If x increases, does y increase? How do you know?*
- *Which variable increases more quickly: x or y? How do you know?*
- *Why is it useful to use the equation of a line to describe how two variables are related?*
- *How did you solve the problem?*

> *Choice 1:* You are interested in planning a pizza party for your school. What sort of sample would you use to decide how many of what sorts of pizza to order?
>
> *Choice 2:* You are interested in planning what sports events to include in a school sports day. What sample would you use to decide which events to include?

CCSS: Statistics & Probability: 7.SP
Mathematical Practice: 3   BIG IDEA: 9.3

The two tasks are similar in difficulty level, and it is more a matter of personal interest as to which might be more appealing to specific students. In both cases, students need to think about what the overall population is, what an appropriate sample size would be, and whether a random or a more organized sample would be more useful.

➤ **Variations.** Sampling related to social justice issues rather than everyday life might be considered instead, for example, finding out how people feel about allowing refugees into the country or offering medical services to people overseas.

> *Choice 1:* Describe what 10 colored cubes you would put in a bag so that the probability of selecting a red one is high but not certain.
>
> *Choice 2:* Describe what 10 colored cubes you would put in a bag so that the probability of selecting a red one is $\frac{2}{5}$.

CCSS: Statistics & Probability: 7.SP
Mathematical Practices: 1, 2, 3   BIG IDEA: 9.4

Before students are comfortable using fractions to describe probability, they are comfortable using qualitative language to describe it (words such as "probably" or a "high probability"). With two choices, students who are not ready to use fractions to describe probability can be successful, and students who can use fractions have the opportunity to do so.

In some ways, ***Choice 1*** can promote richer conversation, even though it may seem to be the simpler task. Students might debate how high is high enough. For example, could there be 8 reds and 2 other colors, or must there be 9 reds? Could there even be 6 reds? Is any probability greater than $\frac{1}{2}$ considered high?

Whichever task was chosen, students could be asked:

- *How many red cubes did you put in your bag?*
- *Why did you choose that number?*
- *Could you have chosen a different number of red cubes?*
- *What colors did the other cubes have to be?*
- *Do the other colors matter?*

➢ ***Variations.*** The task can be adapted by using an alternate qualitative description of a probability for one task and a quantitative description of a probability for the other task. For example, Choice 1 might ask students to fill a bag so that the probability of picking a blue cube is a little higher than the probability of choosing a red cube, while Choice 2 might ask that the probability of picking a blue cube be $\frac{1}{8}$ more than the probability of choosing a red cube.

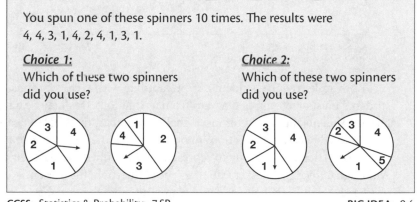

You spun one of these spinners 10 times. The results were
4, 4, 3, 1, 4, 2, 4, 1, 3, 1.

Choice 1:
Which of these two spinners did you use?

Choice 2:
Which of these two spinners did you use?

CCSS: Statistics & Probability: 7.SP   **BIG IDEA:** 9.4
Mathematical Practices: 3, 4

The choice about which spinner is the likely one is somewhat more obvious in ***Choice 1*** than in ***Choice 2***. However, in either case, students must evaluate what happened and recognize that outcomes are not completely predictable, even though they tend to mirror theoretical probabilities to a certain extent.

In the case of ***Choice 1***, it seems highly unlikely that the two most frequent outcomes, 1 and 4, would occur if those were the two smallest sections. It is not certain, however, and students need to realize this. The decision is more difficult in ***Choice 2***. In both cases, the 1 and 4 sections are the biggest ones, as might be anticipated. Even though the fact that there was no result of 5 might dissuade a student from choosing the right-hand spinner in ***Choice 2***, the fact that the 2 section of the right spinner is smaller than the 2 section on the left might make them reconsider.

Students who completed either task could be asked:

- *Which outcomes influenced you the most in deciding which spinner you used?*
- *Can you be sure you are right? Why or why not?*

TEACHING TIP. Students' reasoning skills are fine-tuned more by providing them a choice of which spinner they might have used than by simply asking whether the results make sense for a particular spinner.

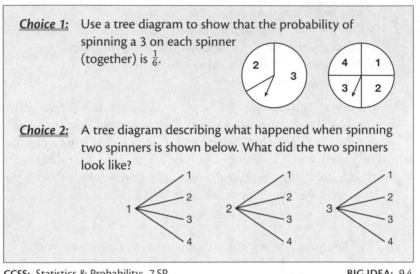

Choice 1:   Use a tree diagram to show that the probability of spinning a 3 on each spinner (together) is $\frac{1}{6}$.

Choice 2:   A tree diagram describing what happened when spinning two spinners is shown below. What did the two spinners look like?

CCSS: Statistics & Probability: 7.SP   **BIG IDEA:** 9.4
Mathematical Practices: 2, 3, 7

Both tasks require students to be familiar with tree diagrams. In *Choice 1*, students must create a tree diagram from a situation. This may be a familiar problem to students, but it is a bit more challenging than the simplest cases they have encountered because the sections of one of the spinners are not of equal size. They will need to determine a way to represent this circumstance in their tree diagram. Although *Choice 2* is a less traditional type of question, the presence of the spinners in *Choice 1* will likely help students figure out what they might do in *Choice 2*.

Whichever choice was selected, a teacher could ask:

- *Why are there two "columns" in the tree diagram?*
- *What do the numbers in the first column represent? In the second column?*
- *How many branches are there? Why does that make sense?*

TEACHING TIP. Creating a situation and then posing a traditional question as one choice and a possible answer to a similar question as another choice often provides good options for parallel tasks. Each choice may serve to provide an element of scaffolding for the alternative choice.

> The probability of an event occurring is $\frac{1}{3}$.
>
> **_Choice 1:_** If you used a die, a spinner, or some other probability
> device, what could the event be and what might the
> device look like?
>
> **_Choice 2:_** If you used two items in any combination of dice,
> spinners, or other probability devices, what could the
> event be and what might the devices look like?

CCSS: Statistics & Probability: 7.SP   **BIG IDEA:** 9.4
      Mathematical Practices: 2, 3

The difference between **_Choices 1_** and **_2_** is whether students can handle only simple events or compound events (i.e., events involving two separate aspects). For example, for **_Choice 1_**, students might simply choose a conventional six-sided die and suggest that the event is rolling a number less than 3; alternately, they might choose a spinner with three equal sections, one of which is red, and the event is spinning a red. For **_Choice 2_**, students must consider combinations. They might begin by thinking about how to write the fraction $\frac{1}{3}$ as a product of two other fractions that are less than 1; these two fractions would then help determine the two devices needed. For example, the two fractions might be $\frac{1}{2}$ and $\frac{2}{3}$, leading the student to choose two spinners, one with two equal sections labeled 1 and 2 and the other with three equal sections labeled 1, 2, and 3, and an event of getting an odd number on both spinners.

Whichever choice was selected, a teacher could ask:

- *Is a probability of $\frac{1}{3}$ likely or unlikely? How do you know?*
- *What device did you choose? What probabilities are easy to describe with it?*
- *How did you decide on your event?*
- *How do you know that the probability of your event is actually $\frac{1}{3}$?*

> A bag contains 3 red cubes, 2 green ones, and 1 blue one.
>
> **_Choice 1:_** You pull a cube out of the bag, record the color, and then
> return it to the bag. Then you pick another cube. What is
> the probability that the cubes are the same color?
>
> **_Choice 2:_** You pull a cube out of the bag, record the color, and do
> not return it to the bag. Then you pick another cube.
> What is the probability that the cubes are the same color?

CCSS: Statistics & Probability: 7.SP   **BIG IDEA:** 9.4
      Mathematical Practices: 3, 5

It might be easier for students to calculate the probability in **_Choice 1_** than in **_Choice 2_**. In **_Choice 1_**, students can draw a traditional tree diagram to observe that the total number of branches is 36 and that there are 14 branches that involve cubes of the same color.

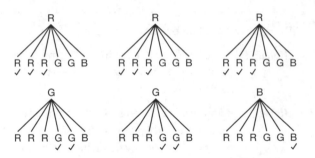

Students may be less sure of how to proceed for **_Choice 2_**. They need to realize that what happens on the second choice depends in part on what happened with the first choice. Their tree diagram is not the normal symmetric type; in this case, the letters for the second choice are different groups each time.

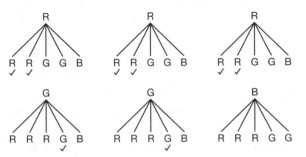

For example, if a red is chosen first, the possible colors for the second pick are 2 reds, 2 greens, and 1 blue, but if a blue is chosen first, the possible colors for the second pick are 3 reds, 2 greens, and 0 blues.

Whichever choice was selected, a teacher could ask:

- *Are you more likely to draw two cubes of the same color or two cubes of different colors? Why?*
- *Is the probability of drawing two cubes of the same color more or less than $\frac{1}{2}$? How do you know?*
- *If you drew a tree diagram to represent your situation, what would it look like?*
- *What is the probability of choosing two cubes of the same color? How do you know?*

Choice 1:   You roll two dice. Is it more likely that the sum is 8 or that the difference is 2?

Choice 2:   You roll two dice. You want an event that is only a bit less likely than rolling a difference of 2. What could it be?

CCSS: Statistics & Probability: 7.SP   **BIG IDEA:** 9.4
       Mathematical Practices: 2, 4, 5

In either case, students must consider the probabilities associated with rolling two dice and might begin by creating a table or chart to list all of the possible outcomes. Some students will be more comfortable with **_Choice 1_**; because the two events are listed, little decisionmaking is required. Other students, who want to be more creative or are more comfortable with ambiguity, might prefer **_Choice 2_**, where they must interpret what "a bit less" means and get to choose the second event.

Whichever choice was selected, a teacher could ask:

- *How likely is it that the two rolls of a die differ by 2?*
- *What would the probability of an event need to be for it to be greater than the probability that the rolls differ by 2?*
- *Is the probability of rolling a sum of 8 less than the probability of rolling a difference of 2?*
- *How did you solve the problem?*

➤ **_Variations._** One variation might be to use other criteria. For example, Choice 1 could be: *Is it more likely that the sum is a prime number or that the difference is not a prime number?*

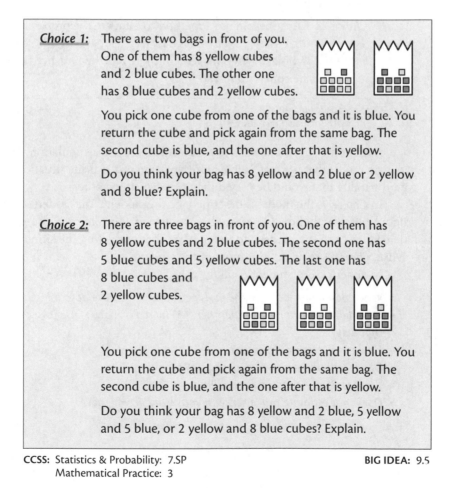

Choice 1: There are two bags in front of you. One of them has 8 yellow cubes and 2 blue cubes. The other one has 8 blue cubes and 2 yellow cubes.

You pick one cube from one of the bags and it is blue. You return the cube and pick again from the same bag. The second cube is blue, and the one after that is yellow.

Do you think your bag has 8 yellow and 2 blue or 2 yellow and 8 blue? Explain.

Choice 2: There are three bags in front of you. One of them has 8 yellow cubes and 2 blue cubes. The second one has 5 blue cubes and 5 yellow cubes. The last one has 8 blue cubes and 2 yellow cubes.

You pick one cube from one of the bags and it is blue. You return the cube and pick again from the same bag. The second cube is blue, and the one after that is yellow.

Do you think your bag has 8 yellow and 2 blue, 5 yellow and 5 blue, or 2 yellow and 8 blue cubes? Explain.

CCSS: Statistics & Probability: 7.SP   **BIG IDEA:** 9.5
       Mathematical Practice: 3

In both instances, students cannot be sure of their answer, but they may be more certain in one case than the other. For students who are less comfortable taking risks, **_Choice 1_** might be a more attractive task.

Parallel Tasks for Grades 6–8

All students could be asked:

- *What color do you think would be picked on a fourth try?*
- *Why do you think that?*
- *Can you be sure?*
- *Suppose you got to choose what colors to put into two or three bags. What colors would you put in the bags so that you could be sure which bag was which once you saw one or two picks? Are there other possible answers, too?*

Choice 1: To estimate the probability that a new baby will be a boy, Tara rolls a die. She says that if she rolls a 1, 2, or 3, she will predict a boy, and if she rolls a 4, 5, or 6, she will predict a girl. Using Tara's method, estimate the probability that, for each of the next 10 days, the first two babies born in Chicago will be boys?

Choice 2: Describe and use a method to estimate the probability that a ball player who is batting 0.250 (getting 1 hit every 4 times at bat) will get 3 or more hits the next five times he comes up to bat?

CCSS: Statistics & Probability: 7.SP   **BIG IDEA:** 9.6
Mathematical Practices: 3, 4, 5

Both of the suggested tasks require students to use a simulation. They involve using a model that is a reasonable representation of a real-life situation to estimate a probability that would be very difficult to determine otherwise.

In *Choice 1*, the model is described for the student. The student then needs to figure out exactly how to set up the experiment. In *Choice 2*, the student must determine both the model and the method for conducting the simulation; in that sense, *Choice 2* requires a bit more of the student.

In either case, a student could be asked:

- *Why does your model for the real-life situation you are describing make sense?*
- *How did you set up your experiment? Why did it make sense to set it up that way?*
- *How many trials did you do?*
- *Do you think there were enough trials?*
- *What is your estimated probability?*
- *Does that estimate make sense to you? Why or why not?*

SUMMING UP

MY OWN QUESTIONS AND TASKS

Lesson Goal: _____   Grade Level: _____

Standard(s) Addressed:

Underlying Big Idea(s):

Open Question(s):

Parallel Tasks:

Choice 1:

Choice 2:

Principles to Keep in Mind:

- All open questions must allow for correct responses at a variety of levels.
- Parallel tasks need to be created with variations that allow struggling students to be successful and proficient students to be challenged.
- Questions and tasks should be constructed in such a way that will allow all students to participate together in follow-up discussions.

The six big ideas that underpin work in statistics and probability were explored in this chapter through about 35 examples of open questions and parallel tasks, as well as variations of them. The instructional examples provided were designed to support differentiated instruction for students in the Grades 6–8 grade band.

The examples presented in this chapter are only a sampling of the many possible questions and tasks that can be used to differentiate instruction in working with statistics and probability. Other questions and tasks can be created by, for example, using alternate probability devices or data distributions. A form such as the one shown here can be a convenient template for creating your own open questions and parallel tasks. Appendix B includes a full-size blank form and tips for using it to design your own teaching materials.

Conclusions

VIRTUALLY EVERY TEACHER of mathematics in the primary, elementary, or middle-school grades is faced with students who vary in mathematical readiness for what is being taught. The differences among students may be cognitive or may relate to learning style or preferences. Regardless of the source or nature of these differences, every teacher wants to help each student succeed in learning. One of the best approaches to fostering this success within today's classroom environment is through differentiating instruction. This book was written to help teachers accomplish this differentiation.

THE NEED FOR MANAGEABLE STRATEGIES

Realistically, teachers do not have the preparation time to develop or the instructional time to deliver alternative programs that specifically target the individual learning levels of the many students with whom they work. What is needed is a way for teachers to engage all students at an appropriate level with a single question or set of related tasks. This book has modeled two core strategies that could be all a teacher requires to meet the mathematical learning needs of most students. These strategies are open questions and parallel tasks.

Open Questions

A question is open when it is framed in such a way that a variety of responses or approaches are possible. To be useful in instruction, it is important that the question focus on an important mathematical concept or skill, that the teacher has considered the readiness of most of his or her students to deal with the question at some level, and that the teacher is comfortable accepting a broad range of appropriate responses from class members.

Parallel Tasks

Two or more tasks are parallel when they are designed to meet the needs of students at different development levels, yet address the same big idea and are close enough in context that they can be discussed simultaneously. In implementing a lesson with parallel tasks, it is important that the teacher be prepared with a series of at least two or three follow-up questions about student work that would be pertinent no matter which task a student had completed. Sometimes the teacher should decide which of the tasks each student is to complete, but at other times the students should be allowed to choose.

DEVELOPING OPEN QUESTIONS AND PARALLEL TASKS

The same general approaches can be applied to developing open questions and parallel tasks for mathematical instruction.

A Three-Step Process

Step 1: Review Examples. This book provides almost 400 open questions and parallel tasks that can serve as models for new questions and tasks. A review of the examples for a particular content strand at a particular grade band will give a good sense of appropriate instructional material. By using the examples, and the variations proposed, a teacher will have a firm foundation upon which to build.

Step 2: Create the Question or Tasks. The form provided in Appendix B can serve as a convenient template for creating new open questions and parallel tasks. Among other things, the form offers a reminder that coherent instruction is served through a focus on big ideas as a means of addressing content standards. Bearing in mind how students differ with respect to specific behaviors related to the targeted big idea, the teacher can create original questions and tasks or adapt them from an existing resource. The objective in all cases should be to produce materials that appropriately challenge a broad range of students, from those who are struggling to those who are "average" to those who are particularly capable.

Step 3: Plan for Follow-Up. The teacher must ensure that the conversation resulting from class work with an open question or set of parallel tasks is inclusive. Careful structuring of questions and coordination of tasks is crucial. In follow-up discussion, the teacher should try to build connections among the various levels of responses that students provide. Making these connections clear will help the struggling student see how the answer that he or she provided can be an important step toward a more sophisticated answer.

Creating an Open Question: An Example

Consider, for example, a situation in which the intent is to teach division of three-digit numbers by one-digit numbers in a classroom where many students are still struggling with their multiplication facts. Those struggling students might not be ready for the types of questions that were initially planned. To adapt the lesson for a differentiated approach, the teacher might choose a straightforward question from a text, such as:

Suppose 4 students were delivering 176 newspapers and decided to share the task fairly. How many papers would each deliver?

Recognizing that some students were not ready for this question, the teacher might ask, instead:

Choose a number of newspapers to be delivered and a number of students to deliver them. The job should be shared fairly. How many papers should each deliver?

The change is simple but important. Now every student can work at a level that benefits him or her. In the discussion of the problem as students share their work, every student in the class (no matter what numbers that person chose) gets to hear how other students handled the situation with larger or smaller numbers. A win-win!

Creating a Set of Parallel Tasks: An Example

Consider, again, the case of the division example posed earlier. It might be that the original problem involving the 176 newspapers is posed, along with an alternate problem in which two students share the work of delivering 24 newspapers. Students might be directed to a specific problem or offered a choice of which problem to complete. However, it is critical that the discussion of the problems be common to both groups. For example, the teacher could ask:

- *What operation did you do to decide how many papers each student would deliver?*
- *Why would you use that operation?*
- *Is there another way you could have determined the answer?*
- *How did you know that each student had to deliver more than 10 papers?*
- *How do you know that each student had to deliver fewer than 100 papers?*
- *How did you figure out how many papers each student had to deliver?*

There are two important points to note: (1) The context being the same is important to making a common discussion possible. (2) It is also important to set up the situation so that there could be common questions beyond simply *What did you do?*

Fail-Safe Strategies

Open questions can be created relatively easily by strategies such as giving an answer and asking for a question for which it is the answer, allowing students to choose numbers within a question, or asking how two items are alike and different.

Parallel tasks can be generated from a single original task by strategies such as changing the complexity of the numbers, shapes, graphs, patterns, equations, or measurements being employed or the complexity of the situations being addressed.

THE BENEFITS OF THESE STRATEGIES

Teachers may at first find it challenging to differentiate their teaching by incorporating open questions and parallel tasks. Before long, however, thinking about ways to open up instruction will become a habit when content that might be too difficult for some students must be taught. The techniques illustrated in this book will become valuable tools for adjusting instruction to accommodate students at all developmental levels.

In the beginning, it might make sense for a teacher to commit to using open questions at least once a week, ultimately aiming to use such questions every one or two class periods. And a teacher might commit to using parallel tasks at least once a week, working toward employing them even more frequently. Eventually, students might even be encouraged to create their own parallel tasks by changing the complexity of situations they are asked to address.

The use of these straightforward but powerful differentiation strategies will lead to a classroom with much higher engagement levels, much less frustration, and much more math learning, thinking, and talking—promoting a much more positive attitude toward math for most students.

Appendix A

Mathematical Content Domains and Mathematical Practices of the Common Core State Standards

MATHEMATICAL CONTENT DOMAINS		
Domain	**Grade Band**	**Chapter**
Counting & Cardinality (CC)	Pre-K–Grade 2, Grades 3–5	1
Number & Operations in Base Ten (NBT)	Pre-K–Grade 2, Grades 3–5	1
Number & Operations—Fractions (NF)	Grades 3–5	2
Ratios & Proportional Relationships (RP)	Grades 6–8	3
The Number System (NS)	Grades 6–8	4
Operations & Algebraic Thinking (OA)	Pre-K–Grade 2, Grades 3–5	5
Expressions & Equations (EE)	Grades 6–8	6
Functions (F)	Grades 6–8	6
Measurement & Data (MD)	Pre-K–Grade 2, Grades 3–5	7
Geometry (G)	Pre-K–Grade 2, Grades 3–5, Grades 6–8	8
Statistics & Probability (SP)	Grades 6–8	9

MATHEMATICAL PRACTICES
1. Making sense of problems and persevering in solving them
2. Reasoning abstractly and quantitatively
3. Constructing viable arguments and critiquing the reasoning of others
4. Modeling with mathematics
5. Using appropriate tools strategically
6. Attending to precision
7. Looking for and making use of structure
8. Looking for and expressing regularity in repeating reasoning

Appendix B
Worksheet for Open Questions and Parallel Tasks

STRATEGIES for developing open questions and parallel tasks are discussed in the introductory chapter and in the Conclusions. The form that appears on the next page can serve as a convenient template for creation of customized materials to support differentiation of instruction in math. The form is also available for download on the Teachers College Press website: www.tcpress.com.

The following fundamental principles should be kept in mind when developing new questions and tasks:

- All open questions must allow for correct responses at a variety of levels.
- Parallel tasks need to be created with variations that allow struggling students to be successful and proficient students to be challenged.
- Questions and tasks should be constructed in such a way that all students can participate together in follow-up discussions.

MY OWN QUESTIONS AND TASKS

Lesson Goal: **Grade Level:** _____

Standard(s) Addressed:

Underlying Big Idea(s):

Open Question(s):

Parallel Tasks:

Choice 1:

Choice 2:

Principles to Keep in Mind:

- All open questions must allow for correct responses at a variety of levels.
- Parallel tasks need to be created with variations that allow struggling students to be successful and proficient students to be challenged.
- Questions and tasks should be constructed in such a way that will allow all students to participate together in follow-up discussions.

Glossary

For each Glossary entry, the chapter and page of first occurrence of the term are given in brackets at the end of the definition.

10-frame. A model with 10 cells, usually 2 rows of 5 columns. [Chap. 1, p. 31]

100 chart. A chart with the numbers from 1 to 100 displayed in order in 10 rows of 10 [Chap. 1, p. 27]

absolute value. The undirected distance of a number on the number line from 0 [Chap. 4, p. 87]

acute triangle. A triangle with three angles less than 90° [Chap. 8, p. 198]

addend. One of the numbers used to create a sum. [Chap. 1, p. 30]

additive relationship. When two numbers are compared by determining their difference, an additive relationship is being used [Chap. 3, p. 77]

algebraic expression. A combination of variables, operation signs, and numbers (e.g., $2n + 4$) [Chap. 6, p. 121]

algorithm. A standard procedure for performing a task (e.g., a standard procedure for adding numbers) [Chap. 1, p. 17]

array. A rectangular arrangement of objects in rows of equal length and columns of equal height [Chap. 5, p. 117]

attribute. A feature of an object (e.g., one of the attributes of a triangle is the relationship between its side lengths) [Chap. 7, p. 141]

attribute blocks. A set of blocks that vary by shape, color, size, and sometimes thickness [Chap. 7, p. 171]

axis. One of the two lines used as a reference in a Cartesian coordinate grid [Chap. 6, p. 128]

bar graph. A way to show and compare data by using horizontal or vertical bars [Chap. 7, p. 141]

base ten blocks. Blocks of different sizes representing place value columns; a larger block is 10 times the size of the next smaller block. Blocks consist of small cubes, rods, flats, and large cubes [Chap. 1, p. 23]

baseline. A vertical or horizontal starting line used so that length measurements can be compared [Chap. 7, p. 143]

BEDMAS. An acronym for: brackets, exponents, division, multiplication, addition, subtraction, indicating the order in which computations in an expression should be performed [Chap. 6, p. 130]

benchmark. Familiar measurements used for comparing other measurements (e.g., 1 inch, 1 foot, the width of a hand, the length of a foot) [Chap. 7, p. 142]

benchmark numbers. Familiar numbers used as referents for other numbers (e.g., 25 is a benchmark for numbers such as 22, 27, or 28) [Introduction, p. 5]

big idea. A statement that is fundamental to mathematical understanding; usually links many specific math outcomes [Introduction, p. 4]

bivariate data. Data designed to show the relationship between two variables [Chap. 9, p. 249]

box plot. A graph designed to show how data are spread out; above a number line the minimum and maximum points are plotted and connected, and the middle 50% of the data are boxed in [Chap. 9, p. 235]

Cartesian coordinate grid. A grid made up of vertical and horizontal intersecting lines; positions on the grid can be located by using two numbers— one describing the horizontal distance from a designated point and one describing the vertical distance from that point [Chap. 6, p. 132]

center of data. A single value that describes an average of a set of data [Chap. 9, p. 235]

center of rotation. The point around which a shape is rotated [Chap. 8, p. 212]

circumference. The distance around a circle [Chap. 8, p. 203]

coefficient. The multiplier for a variable in an equation or expression [Chap. 6, p. 123]

compass. A tool used to draw a circle or part of a circle [Chap. 8, p. 219]

compose. Shapes are composed when they are put together. [Chap. 8, p. 185]

composite numbers. Whole numbers with more than two factors (e.g., 6 is composite because it has factors of 1, 2, 3, and 6) [Chap. 5, p. 116]

compound events. Two or more events that must both be considered in establishing a probability [Chap. 9, p. 235]

concrete graph. A way to show and compare data by using actual objects lined up horizontally or vertically [Chap. 7, p. 151]

congruence. Being of the same size and shape [Chap. 8, p. 186]

conjecture. A hypothesis or guess that something is true [Chap. 1, p. 36]

core. One copy of the smallest part of a repeating pattern that repeats [Chap. 5, p. 109]

cross-section. The two-dimensional shape resulting from cutting through a three-dimensional shape with a flat plane [Chap. 8, p. 186]

cube root. A number that can be multiplied by itself twice to get to a certain value, e.g., 3 is the cube root of 27 since $3 \times 3 \times 3 = 27$. [Chap. 4, p. 87]

Cuisenaire rods. A set of 10 colored rods of lengths 1 cm, 2 cm, 3 cm, . . . , 10 cm [Chap. 5, p. 102]

decompose. Numbers are decomposed when broken into parts, for example, 35 into 30 + 5 or 28 + 7. Shapes are decomposed when cut up into parts. [Chap. 1, p. 18; Chap. 8, p. 185]

denominator. The number on the bottom of a fraction that tells the number of equal parts into which the whole or unit is divided [Chap. 2, p. 53]

density. A number set is called dense if a member of that set can always be found between any two other members of the set. Fractions are a dense set; integers are not [Chap. 2, p. 57]

diagonal. A line segment connecting points in a polygon that are not already connected [Chap. 7, p. 169]

diameter. The distance across a circle through the center [Chap. 8, p. 203]

difference. The result of a subtraction. [Chap. 1, p. 17]

differentiating instruction. Tailoring instruction to meet the needs of different types or levels of students in a classroom [Introduction, p. 1]

dilatation. A transformation that changes the size, but not the shape, of an object [Chap. 8, p. 186]

dividend. The number being shared or grouped in a division situation (e.g., if 9 is divided by 3, 9 is the dividend) [Chap. 1, p. 35]

divisor. The group size or number in a group in a division situation (e.g., if 8 is divided by 2, 2 is the divisor) [Chap. 1, p. 35]

dot plot. A simple histogram-like plot of a small data set where dots, not bars, are used [Chap. 9, p. 235]

double. A multiple of 2 [Chap. 1, p. 41]

double number line. A visual that uses two number lines, in tandem, to show a multiplicative relationship [Chap. 3, p. 80]

dynamic geometry software. A digital tool that allows students to draw geometric shapes and measure them to build and test conjectures [Chap. 8, p. 219]

edge. A line segment formed where two faces of a three-dimensional shape meet [Chap. 8, p. 191]

equation. A statement indicating two values are equal. [Chap. 3, p. 80]

equilateral triangle. A triangle with three equal sides [Chap. 7, p. 149]

equivalent fractions. Two different fractions representing the same part of a whole (e.g., $\frac{1}{4}$ and $\frac{2}{8}$) [Chap. 2, p. 53]

equivalent ratios. Two different ratios representing the same multiplicative comparison. [Chap. 3, p. 77]

experimental probability. The probability that results from an experiment [Chap. 9, p. 235]

exponent. The symbol in a power expression that describes how many times a factor is multiplied by itself (e.g., 4 is the exponent in 3^4, indicating that 3 is multiplied by itself 4 times: $3 \times 3 \times 3 \times 3$) [Chap. 6, p. 123]

exterior angle. An angle outside a shape, formed by extending one side of the shape. [Chap. 8, p. 230]

face. A flat side of a three-dimensional shape (e.g., cubes have six faces that are all squares) [Chap. 8, p. 185]

factor. One of the whole numbers that is multiplied to produce another number (e.g., 4 is a factor of 8, because $2 \times 4 = 8$) [Chap. 1, p. 32]

function. A relationship between two variables where there is one output for each input [Chap. 6, p. 121]

hexagon. A shape with six straight sides [Chap. 2, p. 57]

histogram. A continuous bar graph; bars are joined at the sides because the data they represent are continuous [Chap. 9, p. 235]

hypotenuse. The longest side of a right triangle [Chap. 8, p. 208]

improper fraction. A fraction worth 1 or more. [Chap. 2, p. 53]

increasing pattern. A pattern in which the values keep increasing [Chap. 5, p. 109]

inequality. A statement indicating that one value is greater or less than another [Chap. 6, p. 122]

integer. A counting number, 0, or the opposite of a counting number on the other side of 0 on a number line. [Chap. 4, p. 83]

interior angle. An angle inside a shape. [Chap. 8, p. 230]

inter-quartile range. The difference between the greatest and least value in the middle 50% of a set of data. [Chap. 9, p. 235]

interval. The distance between two endpoints on a graph scale [Chap. 7, p. 180]

irrational number. A number that cannot be written as the quotient of two integers [Chap. 4, p. 83]

isosceles trapezoid. A trapezoid with two nonparallel sides of equal length [Chap. 8, p. 213]

isosceles triangle. A triangle with two sides of equal length [Chap. 7, p. 149]

kite. A quadrilateral with two pairs of equal sides where the equal sides are not parallel [Chap. 7, p. 158]

line of best fit. A line that approximates the relationship shown in a data set relatively well [Chap. 9, p. 249]

line plot. A way to show and compare data that can be sorted into numerical categories; formed by making marks in a column above positions on the number line for each category [Chap. 7, p. 141]

line symmetry. A property of shapes in which one side is a mirror reflection of the other side [Chap. 8, p. 185]

linear association. A relationship between two variables such that a scatterplot of corresponding values forms a line [Chap. 9, p. 240]

linear dimensions. Length measurements associated with a shape [Chap. 2, p. 66]

linear equation. An equation where the variable is raised to the first power, e.g., $4x + 3 = 2$. [Chap. 6, p. 121]

linear growing pattern. A pattern based on the addition of a constant to get from one term to the next [Chap. 6, p. 121]

mean. A value for a set of numbers determined by adding them and dividing by the number of numbers in the set; average [Chap. 7, p. 166]

mean absolute variation. A measure of data spread that is the mean value representing the distances of each piece of data from the mean. [Chap. 9, p. 235]

median. The middle number in an ordered set of numbers [Chap. 9, p. 235]

mental math. Calculating without paper and pencil, using either memory or number relationships. [Chap. 1, p. 27]

mixed number. A number with a whole number and proper fraction part. [Chap. 1, p. 31]

multiple. A number that is the product of two other whole numbers (e.g., 6 is a multiple of 3) [Chap. 1, p. 32]

multiplicative comparison. When two numbers are compared in terms of how many of one fits into the other, the comparison is called multiplicative [Chap. 3, p. 70]

negative integers. The opposites of the counting numbers 1, 2, 3, ... on the other side of 0 on the number line (i.e., –1, –2, –3, ...) [Chap. 4, p. 83]

net. A two-dimensional shape that can be folded to create a three-dimensional object [Chap. 8, p. 186]

nonstandard unit. A measurement used as the basis for describing other measurements, but one that may vary from person to person (e.g., the length of a pencil or the width of a hand might be used as units for describing length) [Chap. 7, p. 141]

number line. A diagram that shows ordered numbers as points on a line [Chap. 1, p. 28]

numerator. The number on the top of a fraction that tells the number of parts of a whole under consideration [Chap. 2, p. 53]

numerical expression. A sequence of numbers joined by operation signs and brackets that would yield a number result [Chap. 4, p. 93]

obtuse triangle. A triangle with one angle greater than 90° [Chap. 8, p. 198]

octagon. An 8-sided shape. [Chap. 8, p. 191]

open questions. Broad-based questions with many appropriate responses or approaches; the questions serve as a vehicle for differentiating instruction if the range of possibilities allows students at different developmental levels to succeed even while responding differently [Introduction, p. 6]

opposites. Two numbers on a number line that are the same distance from 0 (e.g., 8 and –8) [Chap. 4, p. 83]

outlier. A piece of data that does not seem like it "belongs" with the rest of the data. [Chap. 9, p. 238]

pan balance. A tool used to compare weights; when an item is placed on each side of the pan balance, if the pans are at the same level, then the weights are the same [Chap. 5, p. 101]

parallel lines. Two lines that are parallel do not meet [Chap. 8, p. 186]

parallel tasks. Sets of two or more related tasks or activities that explore the same big idea but are designed to suit the needs of students at different developmental levels; the tasks are similar enough in context that they can be considered together in class discussion [Introduction, p. 6]

parallelogram. A quadrilateral with opposite sides that are parallel [Chap. 8, p. 189]

pattern blocks. A set of shapes that includes hexagons, trapezoids, squares, parallelograms, and triangles of particular sizes [Chap. 7, p. 158]

pattern rule. An iterative rule or relationship rule describing how the terms of a pattern are determined (e.g., the pattern 4, 7, 10, 13, ... can be described by the rule "starts at 4 and increases by 3 each time" or by the rule "triple the position number and add 1") [Chap. 5, p. 96]

pentagon. A polygon with five sides [Chap. 8, p. 217]

perfect square. A whole number that is the product of a whole number multiplied by itself. [Chap. 4, p. 89]

perimeter. The distance around a shape [Chap. 6, p. 136]

perpendicular. Two lines or planes are perpendicular if a 90° angle is formed. [Chap. 8, p. 196]

perpendicular distance. The distance from a point to a line that is measured perpendicular to the given line [Chap. 8, p. 196]

pi (π). A number, about 3.14, that represents the ratio of the circumference of a circle to its diameter. [Chap. 4, p. 87]

picture graph. A graph that uses pictures to represent quantities [Chap. 7, p. 141]

place value chart. A chart showing columns of ones, tens, hundreds, thousands, . . . and/or tenths, hundredths, thousandths . . . [Chap. 1, p. 23]

place value system. A system used to represent numbers in which the position of a symbol affects its value (e.g., the 2 in 23 is worth a different amount than the 2 in 217) [Chap. 1, p. 17]

polydrons. Plastic shapes that can be connected to form three-dimensional shapes [Chap. 8, p. 217]

polygon. A closed shape made of only straight sides [Chap. 7, p. 176]

polyhedron. A three-dimensional shape with only polygon faces; the plural is *polyhedra* [Chap. 8, p. 215]

power. A number written in the form a^b, indicating repeated multiplication of a [Chap. 6, p. 130]

power of 10. The number 10 or the result of multiplying 10 by itself repeatedly (e.g., 10, 100, 1000, 10,000) [Chap. 1, p. 50]

prime numbers. Counting numbers with no whole-number factors other than themselves and 1 (e.g., 17 is a prime number because its only factors are 1 and 17) [Chap. 5, p. 102]

prism. A three-dimensional shape with two identical polygon bases that are parallel and opposite each other [Chap. 7, p. 161]

product. The result of multiplication (e.g., in the expression $5 \times 3 = 15$, 15 is the product) [Chap. 1, p. 17]

proper fraction. A fraction less than 1 [Chap. 2, p. 59]

properties. An attribute that applies to all items in a group (e.g., a property of squares is that they have four identical corners) [Chap. 1, p. 47]

proportion. A description of how two numbers are related based on multiplication (e.g., the proportion 25:100 can also be described as the fraction $\frac{1}{4}$, the decimal 0.25, or the ratio 1:4), *or* an equation describing two equal ratios [Chap. 3, p. 69]

proportional thinking. A focus on comparing numbers based on multiplication (e.g., thinking of 12 as four 3s rather than as, e.g., 10 + 2) [Chap. 3, p. 69]

pyramid. A shape with a polygon base and triangular faces that meet at a point [Chap. 8, p. 186]

Pythagorean theorem. A statement that the square of the length of the longest side of a right triangle is the sum of the squares of the lengths of the two other sides [Chap. 8, p. 185]

quadrant. One fourth of a coordinate grid; the first quadrant is to the right and above the center point; the second quadrant is to the left and above the center point; the third quadrant is to the left and below the center point; and the fourth quadrant is to the right and below the center point [Chap. 4, p. 89]

quadratic function. A function of the form $y = a^x + bx + c$ [Chap. 6, p. 127]

quadrilateral. A four-sided polygon [Chap. 8, p. 188]

quotient. The result of division (e.g., in the expression 8 ÷ 4 = 2, 2 is the quotient) [Chap. 2, p. 17]

radius. The distance from the center of a circle to the outer edge [Chap. 8, p. 203]

ratio. A proportional comparison of two or more numbers [Chap. 3, p. 69]

rational numbers. Numbers that can be expressed as the quotient of two integers (e.g., 0.25, –4.5, or $\frac{3}{11}$) [Chap. 4, p. 83]

rectangular prism. A prism with rectangle-shaped bases [Chap. 8, p. 188]

reflection. The mirror image of (or result of flipping) a shape [Chap. 8, p. 186]

reflection line. A flip line [Chap. 8, p. 196]

regular shape. A shape with equal straight sides and equal angles [Chap. 7, p. 178]

rekenrek. A frame of beads featuring 2 or 10 rows, each with 10 beads organized into two groups of 5, each group a different color [Chap. 1, p. 20]

remainder. The amount left after equal groups of a given size have been formed [Chap. 1, p. 35]

repeated addition. An addition where the same number is added to itself one or more times. [Chap. 2, p. 66]

representative sample. A sample that one could argue represents a full population well. [Chap. 9, p. 236]

rhombus. A parallelogram with four sides of equal length [Chap. 7, p. 158]

right angle. A 90° angle [Chap. 7, p. 158]

right triangle. A triangle with a 90° angle [Chap. 8, p. 185]

rotation. A turn [Chap. 8, p. 186]

round. Estimate a number using a benchmark that is generally a multiple of a power of 10 [Chap. 1, p. 17]

scale. The amount that one unit in a graph represents (e.g., if each square in a bar graph represents 5, the scale is 5) [Chap. 6, p. 140]

scale drawing. A drawing of an object or set of objects that is mathematically similar to the real thing [Chap. 8, p. 186]

scalene triangle. A triangle with all sides of unequal lengths [Chap. 8, p. 198]

scatter plot. A set of points plotted on a coordinate grid designed to make relationships between two variables evident [Chap. 9, p. 235]

similarity. A relationship between two shapes such that one is the same shape but of a different size than the other; in effect, an enlargement or reduction [Chap. 8, p. 186]

simplest form. For a fraction, a form in which the numerator and denominator have no factors other than 1 in common [Chap. 2, p. 60]

simulation. A model that can be used to mimic a real-life situation [Chap. 9, p. 235]

simultaneous linear equations. Two or more equations that must hold true at the same time [Chap. 6, p. 121]

skeleton. A representation of a shape that uses only its edges and vertices [Chap. 8, p. 191]

snap cubes. Plastic cubes that can be linked together [Chap. 7, p. 162]

sorting rule. A description of how it is determined which items of a group belong together [Chap. 7, p. 171]

spread of data. A single number to represent the variability of a set of data [Chap. 9, p. 235]

square root. A number that can be multiplied by itself to result in a given number (e.g., the square root of 16 is 4, because $4 \times 4 = 16$) [Chap. 4, p. 87]

standard algorithm. The conventional procedure used for calculation [Chap. 1, p. 46]

standard deviation. A measure of data spread which represents the square root of the average squared distance of data values from the mean [Chap. 9, p. 235]

standard unit. An unambiguous measurement based on units that hold the same meaning for anyone (e.g., inch, mile, square meter) [Chap. 7, p. 141]

straightedge. A tool with a straight edge (like a ruler) without measurements on it [Chap. 8, p. 219]

subitize. Recognize an amount by just looking and not counting [Chap. 1, p. 39]

substitution. Replacement of a variable with a given number [Chap. 6, p. 137]

sum. The result of an addition [Chap. 1, p. 17]

summary statistic. A single number that in some way represents an entire set of data (e.g., mean) [Chap. 9, p. 236]

supplement. Two angles are supplementary if their sum is 180° [Chap. 8, p. 205]

surface area. The total of the area of the surfaces of a three-dimensional object [Chap. 8, p. 202]

table of values. A way to organize related information in a chart with two columns [Chap. 3, p. 75]

tally chart. A set of small lines organized in groups of five to show a count [Chap. 7, p. 153]

tangram. A puzzle in the form of a square made up of seven specific pieces [Chap. 8, p. 193]

theoretical probability. The probability that would be expected when analyzing all of the possible outcomes of a situation [Chap. 9, p. 235]

transformation. A motion that moves a shape from one position to another; translations, reflections, and rotations are common motions studied [Chap. 8, p. 186]

translation. A slide with a vertical and/or horizontal component [Chap. 8, p. 186]

transparent mirror. A mirror that can be seen through; with this type of mirror, a student can draw the image of an object that is viewed in it [Chap. 8, p. 196]

transversal. A line is a transversal to another line if it crosses that line [Chap. 8, p. 206]

trapezoid. A quadrilateral with one pair of parallel sides [Chap. 7, p. 158]

tree diagram. A way to record and count all combinations of events by using lines to form branches [Chap. 9, p. 243]

triangle-based prism. A prism in which the two opposite bases are triangles [Chap. 8, p. 217]

triangle-based pyramid. A pyramid with a triangle base [Chap. 8, p. 210]

two-way tables. A 2 by 2 frequency table that helps the reader relate values of two variables.[Chap. 9, p. 235]

unit. A measurement used as the basis for describing other measurements (e.g., inch is a unit of length) [Chap. 3, p. 69]

unit fractions. Fractions with a numerator of 1 [Chap. 2, p. 53]

univariate data. Data describing a single variable [Chap. 9, p. 235]

variable. A quantity that varies or changes or is unknown; often represented by an open box or a letter [Chap. 3, p. 75]

vertex. A corner of a shape; the plural is *vertices* [Chap. 8, p. 191]

vertical angles. Two angles on opposite sides of where two lines cross [Chap. 8, p. 206]

whole numbers. The numbers 0, 1, 2, 3, . . . [Chap. 1, p. 17]

zero pair. A pair of integers that add to 0, usually +1 and −1. [Chap. 4, p. 85]

zone of proximal development. The range of potential learning that is beyond existing knowledge but that is accessible to a student with adult or colleague support [Introduction, p. 3]

Bibliography

Anderson, M., & Little, D. M. (2004). On the write path: Improving communication in an elementary mathematics classroom. *Teaching Children Mathematics, 10,* 468–472.

Barlow, A. T., & McCrory, M. R. (2011). Engage your students in reasoning and sense making with these effective instructional plans. *Teaching Children Mathematics, 16,* 530–539.

Borgoli, G. M. (2008). Equity for English language learners in mathematics classrooms. *Teaching Children Mathematics, 15,* 185–191.

Bremigan, E. G. (2003). Developing a meaningful understanding of the mean. *Mathematics Teaching in the Middle School, 9,* 22–27.

Bright, G. W., Brewer, W., McClain, K., & Mooney, E. S. (2003). *Navigating through data analysis in Grades 6–8.* Reston, VA: National Council of Teachers of Mathematics.

Carmody, H. G. (2010). Water bottle designs and measures. *Teaching Children Mathematics, 16,* 272–279.

Cavanagh, M., Dacey, L., Findell, C. R., Greenes, C. E., Sheffield, L. J., & Small, M. (2004). *Navigating through number and operations in Prekindergarten–Grade 2.* Reston, VA: National Council of Teachers of Mathematics.

Charles, R. (2004). Big ideas and understandings as the foundation for elementary and middle school mathematics. *Journal of Mathematics Education Leadership, 7,* 9–24.

Clements, D. H. (1999). Geometric and spatial thinking in young children. In J. Copley (Ed.), *Mathematics in the early years* (pp. 66–79). Washington, DC: National Association for the Education of Young Children and National Council of Teachers of Mathematics.

Common Core State Standards Initiative. (2010). Common Core State Standards for Mathematics. Retrieved from http://www.corestandards.org/assets/CCSSI_Math%20 Standards.pdf.

Cuevas, G. J., & Yeatts, K. (2001). *Navigating through algebra, Grades 3–5.* Reston, VA: National Council of Teachers of Mathematics.

Dacey, L., Cavanagh, M., Findell, C. R., Greenes, C. E., Sheffield, L. J., & Small, M. (2003). *Navigating through measurement in Prekindergarten–Grade 2.* Reston, VA: National Council of Teachers of Mathematics.

Dacey, L., & Gartland, K. (2009). *Math for all: Differentiating instruction, Grades 6–8.* Sausalito, CA: Math Solutions Publications.

Dacey, L., & Lynch, J. B. (2007). *Math for all: Differentiating instruction, Grades 3–5.* Sausalito, CA: Math Solutions Publications.

Dacey, L., & Salemi, R. E. (2007). *Math for all: Differentiating instruction, Grades K–2.* Sausalito, CA: Math Solutions Publications.

Danielson, C. (2016). *Which one doesn't belong?* Portland, ME: Stenhouse Publishing.

Drake, J. M., & Barlow, A. T. (2007). Assessing students' levels of understanding multiplication through problem writing. *Teaching Children Mathematics, 14,* 272–277.

Findell, C. R., Small, M., Cavanagh, M., Dacey, L., Greenes, C. E., & Sheffield, L. J. (2001). *Navigating through geometry in Prekindergarten–Grade 2.* Reston, VA: National Council of Teachers of Mathematics.

Forman, E. A. (2003). A sociocultural approach to mathematics reform: Speaking, inscribing, and doing mathematics within communities of practice. In J. Kilpatrick, W. G. Martin, & D. Schifter (Eds.), *A research companion to* Principles and Standards for School Mathematics (pp. 333–352). Reston, VA: National Council of Teachers of Mathematics.

Friel, S., Rachlin, S., & Doyle, D. (2001). *Navigating through algebra, Grades 6–8.* Reston, VA: National Council of Teachers of Mathematics.

Greenes, C. E., Cavanagh, M., Dacey, L., Findell, C. R., & Small, M. (2001). *Navigating through algebra in Prekindergarten–Grade 2.* Reston, VA: National Council of Teachers of Mathematics.

Gregory, G. H., & Chapman, C. (2006). *Differentiated instructional strategies: One size doesn't fit all* (2nd ed.). Thousand Oaks, CA: Corwin Press.

Heacox, D. (2002). *Differentiating instruction in the regular classroom: How to reach and teach all learners, Grades 3–12.* Minneapolis: Free Spirit.

Heacox, D. (2009). *Making differentiation a habit: How to ensure success in academically diverse classrooms.* Minneapolis: Free Spirit.

Imm, K. L., Stylianou, D. A., & Chae, N. (2008). Student representations at the center: Promoting classroom equity. *Mathematics Teaching in the Middle School, 13,* 458–463.

Jones, G. A., & Thornton, C. A. (1993). Children's understanding of place value: A framework for curriculum development and assessment. *Young Children, 48,* 12–18.

Karp, K., & Howell, P. (2004). Building responsibility for learning in students with special needs. *Teaching Children Mathematics, 11,* 118–126.

Kersaint, G. (2007). The learning environment: Its influence on what is learned. In W. G. Martin, M. E. Strutchens, & P. C. Elliott (Eds.), *The learning of mathematics* (pp. 83–96). Reston, VA: National Council of Teachers of Mathematics.

Lovin, A., Kyger, M., & Allsopp, D. (2004). Differentiation for special needs learners. *Teaching Children Mathematics, 11,* 158–167.

Martin, H. (2006). *Differentiated instruction for mathematics.* Portland, ME: Walch.

Mohr, D. J., Walcott, C. Y., & Kastberg, S. E. (2008). Using your inner voice to guide instruction. *Teaching Children Mathematics, 15,* 112–119.

Moyer, P. S. (2001). Using representations to explore perimeter and area. *Teaching Children Mathematics, 8,* 52–59.

Murray, M., & Jorgensen, J. (2007). *The differentiated math classroom: A guide for teachers, K–8.* Portsmouth, NH: Heinemann.

Myller, R. (1991). *How big is a foot?* New York: Random House.

National Council of Teachers of Mathematics. (2000). *Principles and standards for school mathematics.* Reston, VA: National Council of Teachers of Mathematics.

National Council of Teachers of Mathematics. (2014). *Principles to actions: Ensuring mathematical success for all.* Reston, VA: National Council of Teachers of Mathematics.

O'Connell, S. (2007a). *Introduction to communication, Grades PreK–2.* Portsmouth, NH: Heinemann.

O'Connell, S. (2007b). *Introduction to communication, Grades 3–5.* Portsmouth, NH: Heinemann.

Pugalee, D. K., Frykholm, J., Johnson, A., Slovin, H., Malloy, C., & Preston, R. (2002). *Navigating through geometry in Grades 6–8.* Reston, VA: National Council of Teachers of Mathematics.

Sakshaug, L. (2000). Which graph is which? *Teaching Children Mathematics, 6,* 454–455.

Schifter, D., Bastable, V., & Russell, S. I. (1997). Attention to mathematical thinking: Teaching to the big ideas. In S. Friel & G. Bright (Eds.), *Reflecting on our work: NSF teacher enhancement in mathematics K–6* (pp. 255–261). Washington, DC: University Press of America.

Sheffield, L. J. (2003). *Extending the challenge in mathematics: Developing mathematical promise in K–8 students.* Thousand Oaks, CA: Corwin Press.

Sheffield, L. J., Cavanagh, M., Dacey, L., Findell, C. R., Greenes, C. E., & Small, M. (2002). *Navigating through data analysis and probability in Prekindergarten–Grade 2.* Reston, VA: National Council of Teachers of Mathematics.

Small, M. (2005a). *PRIME: Professional resources and instruction for mathematics educators: Number and operations.* Toronto: Thomson Nelson.

Small, M. (2005b). *PRIME: Professional resources and instruction for mathematics educators: Patterns and algebra.* Toronto: Thomson Nelson.

Small, M. (2006). *PRIME: Professional resources and instruction for mathematics educators: Data management and probability.* Toronto: Thomson Nelson.

Small, M. (2007). *PRIME: Professional resources and instruction for mathematics educators: Geometry.* Toronto: Thomson Nelson.

Small, M. (2010a). Beyond one right answer. *Educational Leadership, 68,* 28–32.

Small, M. (2010b). *PRIME: Professional resources and instruction for mathematics educators: Measurement.* Toronto: Nelson Education Ltd.

Small, M. (2017). *Making math meaningful to Canadian students: K–8* (3rd ed.). Toronto: Nelson Education Ltd.

Small, M., & Lin, A. (2010). *More good questions: Great ways to differentiate secondary mathematics instruction.* New York: Teachers College Press; co-published with NCTM.

Small, M., et al. (2011a). *Leaps and bounds toward math understanding: Teacher's resource 3/4.* Toronto: Nelson Education Ltd.

Small, M., et al. (2011b). *Leaps and bounds toward math understanding: Teacher's resource 5/6.* Toronto: Nelson Education Ltd.

Tomlinson, C. A. (1999). *The differentiated classroom: Responding to the needs of all learners.* Alexandria, VA: Association for Supervision and Curriculum Development.

Tomlinson, C. A. (2001). *How to differentiate instruction in a mixed ability classroom* (2nd ed.). Alexandria, VA: Association for Supervision and Curriculum Development.

Tomlinson, C. A., & McTighe, J. (2006). *Integrating differentiated instruction and understanding by design.* Alexandria, VA: Association for Supervision and Curriculum Development.

Tompert, A. (1990). *Grandfather Tang's story.* New York: Crown.

Vygotsky, L. S. (1978). *Mind in society: The development of higher psychological processes.* Cambridge, MA: Harvard University Press.

Westphal, L. (2007). *Differentiating instruction with menus: Math.* Austin, TX: Prufrock Press.

Williams, L. (2008). Tiering and scaffolding: Two strategies for providing access to important mathematics. *Teaching Children Mathematics, 14,* 324–330.

Index

The Index is divided into two sections: the Index of Subjects and Cited Authors and the Index of Big Ideas. The Index of Subjects and Cited Authors covers instructional concepts as well as listings for content areas and grade bands; the names of authors cited in the text are also included. The Index of Big Ideas covers mathematical content broken down by big idea, providing access to the broad concepts presented. Individual mathematical terms are not indexed. The Glossary (pages 267–276) lists the primary mathematical terms featured in the text and instructional examples. Each Glossary entry ends with a chapter and page designator identifying the location of the first occurrence of the term.

INDEX OF SUBJECTS AND CITED AUTHORS

INDEX OF BIG IDEAS

Counting & Cardinality and Number & Operations in Base Ten

About the Author

MARIAN SMALL is the former dean of education at the University of New Brunswick. She speaks regularly about K–12 mathematics instruction.

She has been an author on many mathematics text series at both the elementary and the secondary levels. She has served on the author team for the National Council of Teachers of Mathematics (NCTM) Navigation series (pre-K–2), as the NCTM representative on the Mathcounts question writing committee for middle school mathematics competitions throughout the United States, and as a member of the editorial panel for the NCTM 2011 yearbook on motivation and disposition.

Dr. Small is probably best known for her books *Good Questions: Great Ways to Differentiate Mathematics Instruction* and *More Good Questions: Great Ways to Differentiate Secondary Mathematics Instruction* (with Amy Lin). *Eyes on Math: A Visual Approach to Teaching Math Concepts* was published in 2013, as was *Uncomplicating Fractions to Meet Common Core Standards in Math, K–7*. She authored *Uncomplicating Algebra to Meet Common Core Standards in Math, K–8* in 2014 and *Teaching Mathematical Thinking: Tasks and Questions to Strengthen Practices and Processes* in 2017. She is also author of the three editions of a text for university preservice teachers and practicing teachers, *Making Math Meaningful to Canadian Students: K–8*, as well as the professional resources *Big Ideas from Dr. Small: Grades 4–8*; *Big Ideas from Dr. Small: Grades K–3*; and *Leaps and Bounds toward Math Understanding: Grades 3–4, Grades 5–6,* and *Grades 7–8,* all published by Nelson Education Ltd. More recently, she has authored a number of additional resources focused on open questions, including *Open Questions for the Three-Part Lesson, K–3: Numeration and Number Sense; Open Questions for the Three-Part Lesson, K–3: Measurement and Pattern and Algebra; Open Questions for the Three-Part Lesson, 4–8: Numeration and Number Sense;* and *Open Questions for the Three-Part Lesson, 4–8: Measurement and Pattern and Algebra,* all published by Rubicon Publishing.

She also led the research resulting in the creation of maps describing student mathematical development in each of the five NCTM mathematical strands for the K–8 levels and has created the associated professional development program, PRIME. She is currently working with Rubicon Publishing to develop a K–8 digital teaching resource, *MathUp,* that assists teachers to deliver a math program that prioritizes the development of students as mathematical thinkers.